In Memory of

Raymond & Alice Winther

DISTRIBUTED BY
HARRY N. ABRAMS, INC.
PUBLISHERS

1000 YEARS
OF RUSSIAN GEMS
AND JEWELS

KREMLIN GOLD

THE STATE MUSEUMS OF THE MOSCOW KREMLIN

THE HOUSTON MUSEUM OF NATURAL SCIENCE

THE FIELD MUSEUM

CONTENTS

FOREWORD

DIRECTORS' FOREWORD

Truett Latimer, President
Houston Museum of Natural Science

John W. McCarter, Jr., President and CEO
The Field Museum

The book you hold in your hands, and the stunning exhibition it accompanies, represent the fruits of a unique, challenging, and immensely rewarding endeavor among three new partners in scientific and cultural exchange.

The Houston Museum of Natural Science and Chicago's Field Museum, in collaboration with the Moscow Kremlin museums, have organized and are exclusively presenting an unprecedented exhibition: *Kremlin Gold: 1000 Years of Russian Gems and Jewels.*

This exhibition features over one hundred archaeological relics, objects of fine and decorative arts, and breathtaking jewelry drawn exclusively from the collections of the Moscow Kremlin Armory Palace. Many of these objects have never been viewed publicly in the United States. More than a simple blend of minerals and metallurgy, these glittering works provide a unique view of the cultural, historical, political, and religious character of Russia and its people over the last millennium. Many objects in this exhibition are associated with the storied names from Russia's past: Ivan the Great, Ivan the Terrible, Boris Godunov, Peter the Great, Catherine the Great, Nicholas and Alexandra.

Here on the threshold of a new millennium, *Kremlin Gold* illuminates the thousand-year epic of Russian history in a way that hasn't been told to the American public until now. While these objects on their own exemplify artistic and technical mastery, they also serve as signature embodiments of Russian culture.

With the break-up of the Soviet Union, new opportunities for American audiences to discover Russia and its history have become possible. These works, both sacred and secular, historical and contemporary, reflect the creativity of the Russian people as well as the varied epochs and social systems of the past. And, just as compelling, they let us see the continuance of great traditions.

Beyond the inherent beauty of the objects themselves and the dramatic tales they tell, however, *Kremlin Gold* represents something more. The project was initiated in 1993 when Truett Latimer arranged a meeting in Moscow with Dr. Irina Rodimtseva, the director of the Moscow Kremlin's State Historical-Cultural Museum Preserve. Impressed with the immeasurable cultural riches held in stewardship by the Kremlin Museums, during this meeting Truett secured an official Protocol of Intent from Kremlin officials in which they agreed to pursue an exclusive exhibition of Kremlin treasures in Houston. Truett was joined by his Curator of Gems and Minerals Joel Bartsch and, over the next five years, their negotiations with Kremlin officials resulted in the signing of a cooperative agreement that would benefit both institutions.

Realizing that an exhibition of this nature would be a perfect extension of Houston's own fine collections of gems and minerals, Truett also knew that a project of this magnitude would be better served through collaboration with another noted U.S. institution. With John McCarter and The Field Museum, whose explicit mission is to explore the world and its peoples while promoting active cultural exchange, the perfect creative partnership was born. The insights into anthropology and cultural history that John's team has brought to the project have significantly enriched the *Kremlin Gold* story.

Staff at both institutions worked hand-in-hand to make this challenging conception a reality. Teaming up on research visits to Moscow, exchanging curatorial expertise and administrative skills, and sharing design inspirations, the project leaders from each side of this partnership have provided an inspiring model of creative collaboration.

Together the Houston Museum of Natural Science and The Field Museum have taken the first steps into what we feel is a model for museum exhibitions of the twenty-first century. Challenging, fruitful collaborations between American and international institutions will enable us to explore and present the world's scientific and cultural riches and bring down the barriers that have so often separated peoples in the past. Our partnership with our new friends in Russia, whose assistance and openness has been truly extraordinary, has brought both professional and personal rewards. If visitors to this exhibition and readers of this catalogue begin to see Russia with the fresh understanding that we now enjoy, the wisdom of this collaboration will have been proven.

On behalf of the Houston Museum of Natural Science and The Field Museum, we invite you to explore and enjoy *Kremlin Gold: 1000 Years of Russian Gems and Jewels.*

DIRECTOR'S FOREWORD

Irina I. Rodimtseva, Director
State Historical-Cultural
Museum Preserve, Moscow Kremlin

FOREWORD

Kremlin Gold: 1000 Years of Russian Gems and Jewels is more than a unique collection of priceless works of art. It is a true cultural treasure as well.

Part of a twenty-year-long program through which the Moscow Kremlin has shared its resources with American museums, this exhibition is noteworthy for the broad sweep of time it encompasses – the entire passing millennium. It includes pieces from the Russian state treasury (The Armory of the Moscow Kremlin), religious objects, and secular ornaments of all kinds, all fashioned from precious metals and gems with consummate skill and artistry. Many of these objects are leaving the country for the first time.

These spectacular pieces were crafted by the best artisans of Russia's prominent jewelry centers, workshops, factories, and firms. All are connected with important people and events in Russian history. Through these extraordinary objects, visitors to the exhibition can trace the history of Russian jewelry work from its most ancient forms to its latest innovations, and explore its ties to the art of other countries.

The gathering of so many precious objects from the Black Sea region, Central Russia, and from the Moscow Kremlin itself is unprecedented. From the remarkable products of fifteenth-century Russian goldsmiths to later examples from the Kremlin workshops, high points in the history of Russian gold and silver work and jewelry making are represented here. Although European styles exerted a strong influence on art and craftsmanship, especially in the eighteenth century, Russian traditions have endured. The gems, the jewels, the regalia of the tsars, and the priceless heirlooms of the church presented in *Kremlin Gold* all reflect these strong and lasting ties to the national heritage.

Many of the artists whose work is included in the exhibition are well known beyond the boundaries of Russia. They include Paul Ovchinnikov, Ivan Khlebnikov, Orest Kurliukov, Nicholas Nemirov-Kolodkin, and, of course, the firm of Carl Fabergé. The name Fabergé has come to be synonymous with technical innovation, performance of the highest quality, and a remarkable artistic imagination in jewel work.

Contemporary works in the exhibition include exclusive factory-made jewelry and handmade originals that confirm that the Russian pride in craftsmanship and high artistry endures and continues to develop, as it has for so many centuries. *Kremlin Gold: 1000 Years of Russian Gems and Jewels*, which opens on the historic threshold of the millennium, celebrates those traditions while looking forward to the cultural flowering of the next thousand years.

PREFACE

CURATOR'S PREFACE

Emma P. Chernukha, Curator
The State Historical-Cultural
Museum Preserve, Moscow Kremlin

The artworks in this catalogue are organized chronologically to present the spectrum of Russian jewelry and metal arts over the last one thousand years and demonstrate the continuity of Russia's artistic traditions from the distant past to the present day. The essays and an historical timeline illuminate the growth of Moscow's Kremlin from a provincial stronghold of earthen walls and oak palisades to the grand architectural ensemble that greets the visitor today. They also illustrate Moscow's role as the center of Russian political and cultural life. Selected objects from the exhibition appear in the timeline to provide a frame of reference for the reader. Two supplements at the back of the book, presenting the mineralogy of Russia and a geographic description of the land itself, describe the extraordinary wealth of natural resources that made this art possible and shaped Russia's growth as a nation.

In tracing the development of Russia's jewelry crafts over this vast sweep of history, distinct periods, trends, and styles emerge. In the catalogue, attention is paid to the origins of Russia's jewelry arts in the craft traditions of the northern Black Sea coastal regions, which pre-date our thousand-year spread by almost five centuries; to the works of Russian masters before the Mongol invasions of the thirteenth century; and to the splendid relics in the tsars' treasuries, produced by the workshops of the Moscow Kremlin between the fifteenth and the seventeenth centuries.

The eighteenth century is represented by objects from the various artistic centers of imperial Russia, the nineteenth by works from the premier jewelry houses in both Moscow and St. Petersburg. No such exploration is complete without a look at the work of Carl Fabergé, the outstanding jewelry legend of the early twentieth century. Objects from the second half of the twentieth century illustrate the resurgence of jewelry workshops and factories and the emergence of new generations of celebrated Russian designers and craftsmen.

The object descriptions accompanying the plates identify both the creators and original owners of the artwork, where these are known, as well as an explanation of an object's function and the materials and technique behind its production. The first use of a Russian term in any given entry appears in italics. The headnote accompanying each plate identifies the place and time of completion, the materials used, and the technical process employed. The width, height, or length of an object is provided to indicate scale.

ACKNOWLEDGMENTS

CURATORS

Emma P. Chernukha
Curator, State Historical-Cultural Museum Preserve, Moscow Kremlin

Joel A. Bartsch
Curator of Gems and Minerals, the Houston Museum of Natural Science

EXHIBITION PROJECT TEAM

STATE HISTORICAL-CULTURAL MUSEUM PRESERVE, MOSCOW KREMLIN

The exhibition was organized with the participation and support of the employees of The State Historical-Cultural Museum Preserve, Moscow Kremlin

I.A. Rodimtseva, *Director*

N.S. Vladimirskaya, *Deputy Director*

O.I. Mironova, *Deputy Director*

A.K. Levykin, *Head of the Armory Ward*

E.P. Chernukha, *Exhibition Concept and Content, Catalogue Compilation*

I.A. Rodimtseva, M.V. Martynova, I.D. Kostina, *Essays*

Object descriptions provided by:

T.D. Avdusina [nos. 6–11]

I.A. Bobrovnitskaya [nos. 19–22, 44–47]

I. I. Vishnevskaya [nos. 23, 28, 27]

I.V. Gorbatova [nos. 116–122]

S.Y. Kovarskaya [nos. 71,73–84, 88]

I.D. Kostina [nos. 55–70, 72]

M.V. Martynova [nos. 1–5, 12–14, 16–18, 24, 25]

E.A. Morshakova [nos. 15, 41]

T.N. Muntyan [nos. 85–87, 89–93, 95–101, 105–111, 113–115, 123]

V.M. Nikitina [no. 94]

L.N. Peshekhonova [nos. 124–140]

E.A. Yablonskaya [no. 26]

N.S. Vladimirskaya, *Scientific editor*

E.B. Gusarova, I.A. Sterligova, *Russian text editors*

N.N. Alekseev, V.A. Seregin, Milar Company, Vienna, *Transparencies*

HOUSTON MUSEUM OF NATURAL SCIENCE

Joel A. Bartsch, *Curator of Gems and Minerals*

Lisa Rebori, *Manager of Collections and Registration*

Hayden Valdes, *Director of Exhibitions*

Lex Vanderende, *Conceptual Design*

Mark Jirczik, Ernesto Perez, *Design, Fabrication, & Installation*

THE FIELD MUSEUM

Laura Gates, *Vice President, International*

Sophia Shaw, *Director of Exhibitions*

Abigail Sinwell, *Manager, Temporary Exhibitions*

David Foster, *Senior Coordinator for Temporary Exhibitions*

David Layman, *Exhibition Design*

Lori Walsh, *Graphic Design*

Matt Matcuk, *Exhibit Development*

Nel Fetherling, *Production Supervisor*

RESEARCH AND TRANSLATION

Julia Khachatryan
Michael Wasserman
Dale Pesman
Emma Krasov

CATALOGUE CONCEPT, DESIGN, AND PRODUCTION

studio blue, *Chicago, Illinois*

PHOTOGRAPHY

Thomas DuBrock, *Houston, Texas*

Jeff Scovil, *Phoenix, Arizona*

Harold and Erica Van Pelt, *Los Angeles, California*

EDITORIAL ASSISTANCE

Joanne Trestrail
Tony Hodgin
George Paterson
Paul Bernhard

ASSISTANCE WITH INTERNATIONAL TRAVEL PROVIDED BY BRITISH AIRWAYS

HISTORY

THE MOSCOW KREMLIN AND THE ARMORY MUSEUM

Irina I. Rodimtseva, Director
State Historical-Cultural
Museum Preserve, Moscow Kremlin

The Moscow Kremlin is the heart of Russia's capital. It is not surprising that to many people around the world, the words "Moscow" and "Kremlin" are nearly synonymous, since the Kremlin is not only the historical center of Moscow, but the nucleus around which the Russian state was formed.

In the course of its 800-year history, the Kremlin has been the setting for many dramatic events, the memory of which lives on in this unique ensemble of buildings. The Kremlin has been both a witness to Russian history and a participant. Here grand dukes and autocrats held the reins of the Russian empire, metropolitans and patriarchs lived and feasted in splendor, invading armies were repulsed, and revolutions rose and fell. Many of the legendary names from Russia's past lived inside the Kremlin walls or made their mark here: Yuri Dolgoruky, Dmitry Donskoy, Ivan the Great, Ivan the Terrible, Boris Godunov, Peter the Great. History continues in the Kremlin today, since it houses the residence of the President of the Russian Federation. The ancient "city-castle" on Borovitsky Hill, built and rebuilt many times, is truly a national treasure and a sacred place, the heart of the Russian nation and the spiritual center of the Russian Orthodox religion.

THE KREMLIN

Tradition sets the date of the Moscow Kremlin's founding in 1147, when prince Yuri Dolgoruky hosted his allies on the hillside site above the Moscow and Neglinnaya Rivers that would eventually become the center of the Russian nation. The Russian word *kremlin* means fort or stronghold; in the turbulent, often violent days of Old Russia, many cities were protected by their own kremlin. Chronicles record that, within several years of his ceremonial feast, Yuri began construction of a kremlin at Moscow, fortifying the heart of the settlement with earthworks and a palisade. Excavations on Borovitsky Hill in 1956–60, during the construction of the Palace of Congress, revealed archaeological remains that supported the evidence of the chronicles' tale.

During the twelfth and thirteenth centuries, Moscow grew in size and power and its trade and crafts flourished. Its prominence as a trading center attracted commerce from a wide radius: glass bracelets from Kiev, coins from Central Asia and Armenia, copper from the Volga regions, boxwood from the Caucasus, and amphoras of wine and spices from the Black Sea attest to the importance of the city.

As Moscow grew in status, it also became more vulnerable to attack and siege. During the bitter feuds for supremacy among rival princes that plagued Old Russia, the Kremlin was simultaneously the prince's residence, the center of trade, and the people's shelter in times of war. Mongol-Tatar incursions of the thirteenth and fourteenth centuries repeatedly targeted Moscow; time and again the wooden city and its Kremlin were burned to the ground.

Moscow's ascendancy continued, however, and its position was solidified in the mid-fourteenth century by Ivan I (Kalita or "Moneybags"), a shrewd prince who allied himself with the Tatars against his Russian competitors, took over the right to collect tribute on their behalf, and laid the foundation of Moscow's future dominance. During Ivan's reign, Moscow doubled in size: massive walls of oak were built to enclose the citadel and Ivan, who possessed a strong sense of tradition, rebuilt many of Moscow's churches on the site of earlier structures. Ivan graced the Kremlin precinct with the Cathedral of the Assumption, based on prototypes in Vladimir; the Church of the Savior, a private chapel for the prince; and the Cathedral of the Archangel Michael, where Moscow's princes and tsars would be buried for the next three hundred years. When the patriarch of the Russian Orthodox church took up residence in Moscow, the city's status as the leader of the nation was assured.

Although Ivan's churches were built of stone, Moscow was still essentially a wooden city that succumbed repeatedly to disastrous fires. In 1367, however, prince Dmitry Donskoy began to build the stone walls that marked the first stage of the impregnable Kremlin of later centuries. Dmitry's stone fortress, a great rarity in its time, extended the Kremlin's area to approximately its present size and indeed proved able to withstand assault by both Lithuanian invaders and the Tatar armies.

Finally, in the last quarter of the fifteenth century, prince Ivan III (the Great) raised the red brick Kremlin walls that we know today. Ivan, a towering figure in Russian history, greatly expanded Moscow's territory and set the stage for the growth of the empire and Russian involvement in European affairs. Ivan ended Russia's isolation from the west when he married the Byzantine princess Sophia Paleologue, thus adding the heritage of Byzantium. Sophia herself may have been responsible for bringing Italian architects such as Aristotle Fioravanti to Moscow. These masters applied Italian Renaissance ideas to their renovations of the Kremlin's ancient churches and palaces.

The modern Kremlin's walls and towers, therefore, were designed by Italian and Russian architects in the fifteenth century. The twenty-one towers in the perimeter were completed about 200 years later, with gates passing through the tallest and most massive of them. Until 1934 they were topped with the state emblem, a two-headed eagle. Today lighted stars top the steeples of the Spasskaya, Nikolskaya, Troitskaya, Borovitskaya, and Vodozvodnaya towers, and many of Moscow's ancient streets and avenues are oriented toward the main towers of the Kremlin.

The ancient arrangement of the Kremlin's buildings remains intact. Along the edge of Borovitsky Hill, with its main facade facing the Moscow River, is the Grand Kremlin Palace, actually a conglomerate of palaces built over the past five centuries. In the heart of the Kremlin itself, Cathedral Square forms the epicenter of the Russian Orthodox Church. Here the Assumption Cathedral (1475–79), the Annunciation Cathedral (1484–89), the Church of the

Archangel (1505–8), Ivan the Great's Bell Tower with the Assumption Belfry (1505–8), and the Patriarch's Palace (seventeenth century) combine in a magnificent ensemble where the pageantry of history is almost palpable. In the nearby Senate and Arsenal, governmental functions are carried out. The majestic building of the Senate (1776–88), with its glowing yellow walls and classical architectural details, was designed by the architect M. F. Kazakov. Following recent renovations, it has served as the residence of the President of the Russian Federation.

THE ARMORY MUSEUM

Inseparably linked to the history of Russia and to the Kremlin itself, the Armory comprises a virtual encyclopedia of Russian art and history. Russia's oldest museum, which in 1991 became the foundation of the State Museum: Moscow Kremlin, the Armory houses works associated with the famous names and figures from Russian history. Here can be found the richest collection in Russia of works of Russian and foreign decorative arts from the fourth to the twentieth centuries, including arms and armor, state regalia, ceremonial dress, gold and silverware, carriages, liturgical objects and vestments, and diplomatic gifts from all over the world. The splendor of these pieces attests to the power of the grand princes and tsars and of the Russian state.

The museum takes its name from the historic armory, the Kremlin workshops that produced and stored arms and armor for the tsar and his troops. For centuries, the large households of the princes and tsars were supplied with everything they needed – from dishes and jewelry to weapons and ceremonial regalia – by the Kremlin armory workshops. Many of the objects were produced for ceremonial and ritual use, as symbols of power and national pride. By the sixteenth century the tsar's treasury was already engaging foreign visitors, whose journals and letters express amazement at the magnificence of the Russian court. The special position of the Armory among Russian museums, in fact, derives from its status as keeper of the holy regalia of the highest governmental power: wreaths, crowns, scepters, orbs, thrones, and coronation clothing. The Kremlin's cathedrals made an equally memorable impression with the richness of clerical attire and ceremonial objects on view there.

Although an official armory is mentioned in manuscripts dating from the early sixteenth century, its roots almost certainly reach back to the early days of the Moscow Kremlin, when princes like Yuri Dolgoruky and Dmitry Donskoy were solidifying the growing power of the young Russian state. Many of the oldest pieces in the museum's collections found their way to Moscow during this period from various cities of Old Russia, such as Kiev, Novgorod, Yaroslavl, Kostroma, Chernigov, and Ryazan, or from Byzantium, the southern Slavic regions, Georgia, and the northern coast of the Black Sea.

The beginnings of a true collection, however, lie in the fourteenth and fifteenth centuries, when the Russian state began to be centralized around Moscow. The future Armory collections began to coalesce at this time, as treasuries were assembled from works of great historical and artistic value. This period saw the birth of the Grand Prince's Treasury, a special chamber in the Moscow Kremlin that was housed in various locations throughout the Kremlin over the centuries. Inventories and records cite fabled rulers whose treasures were kept here: Ivan I (Kalita or "Moneybags"), Dmitri Donskoy, Ivan III (the Great), Ivan IV (the Terrible), Boris Godunov, and others.

By the sixteenth century the office of the tsar's Armorer had been established and documents began to record the activity of the Kremlin workshops. The collections suffered greatly from fires and political upheaval during the "Time of Troubles" (1605–13), as metalwork was melted down and minted into coin for invading Polish and Swedish mercenaries and false claimants to the throne plundered the treasuries and sold off the valuables. The establishment of the Romanov dynasty in 1613, however, saw the fortunes of the Kremlin workshops begin to wax again. Numerous specialized workshops arose, among them the Equestrian Department, the Tsarina's Workshop and Silver Department, and the Goldsmith's Department.

During the seventeenth-century reigns of Romanov tsars Mikhail I and Alexis I, the Kremlin workshops reached their apogee of production. Countless lavish works of decorative arts from this period, many of them produced by Western European artisans or influenced by European styles, illustrate the immense wealth and splendor of the Russian tsars and their courts. In 1712, however, when Peter the Great moved Russia's capital to St. Petersburg, the Kremlin workshops were closed down and the treasure depositories merged into one entity, known as "the Workshops and the Armory." Peter deposited in the Armory much of the spoils captured from the Swedes during the Great Northern War (1700-21) and, in 1718, mounted an exhibition of hereditary relics such as tsars' crowns and garments – the first "museum exhibition" of Kremlin Armory material.

The priceless collections of Kremlin artifacts found a permanent home in 1851, when the present Armory building was completed. As part of the Grand Kremlin Palace construction project decreed by Tsar Nicholas I, who reigned from 1825 to 1855, architect Konstantin A. Thon integrated many of the ancient palaces and churches of the Kremlin precinct into one grand, unified ensemble. At a time when many of the royal and imperial treasuries throughout Europe were being transformed into public institutions, the new Armory was designed specifically as a museum. Thon used rich yellow for the facades and crisp white classical details to make the Armory, the Arsenal, and the Grand Kremlin Palace the premier examples of nineteenth-century Russian architecture in Moscow.

Over the centuries, the Armory's sacred relics and treasures have repeatedly had to be saved from the threat of destruction by outsiders, from nomadic invaders to Napoleon's troops. Withstanding such attacks gave the Kremlin and the Armory the reputation of a reliable shelter. Even today, nearly 800 cannons, veterans of many of Russia's greatest battles, remain at the foundation of the Arsenal building.

The turbulent twentieth century brought new threats. During World War I, treasures of the Russian empire were brought from St. Petersburg, Warsaw, and other imperial centers and hidden behind the Kremlin walls. During the Bolshevik revolution of October, 1917, with the Kremlin under heavy fire, the curators of the Armory did not leave their posts but stayed to protect the Kremlin's collections. In the 1920s, as countless works of art were lost all across Russia, the Armory was one of the few places of safekeeping. In the 1930s, many important works of art were sold abroad, which dealt a serious blow to the state heritage. Finally, during World War II, the threat of Moscow's capture by Hitler's armies forced the evacuation of the Armory's collections to the safety of the Ural mountains.

In the 1950s the Moscow Kremlin and Armory opened to the general public. Twenty years later, a careful restoration of the walls, towers, buildings, and palaces of the Kremlin was carried out, with many important archaeological discoveries on Borovitskiy Hill made in the process. In the Armory, a permanent exhibition, the Diamond Fund of the USSR, was inaugurated. Starting in the 1990s, the Patriarch of Moscow and all Russia, Alexei II, has regularly led festive masses in the Kremlin cathedrals. Presently, as many as three million people visit the Moscow Kremlin every year, with thousands attending plays and other events in the State Kremlin Palace.

Modern visitors marvel at the richness of history represented by the Kremlin's ancient artifacts as well as by the best pieces of contemporary jewelers. They are fascinated not only by the largest works but also by small masterpieces such as the exquisite Easter egg made by the firm of Carl Fabergé to commemorate the construction of the Trans-Siberian Railway in 1900.

One of UNESCO's Monuments of International Cultural and Natural Heritage, the Moscow Kremlin is among the most famous and frequently visited museums in the world. It is a sacred place of Russia and also a priceless contribution to world culture.

A TIMELINE OF RUSSIAN POLITICAL HISTORY

Emma Krasov

For many centuries before the rise of Moscow and the creation of a unified nation, Russia was a patchwork of many colors, with a rich diversity of cultural and ethnic threads. Slav, Norseman, Byzantine, Central Asian nomad – all left their mark as the lands of ancient *Rus* witnessed the shifting tides of migrating tribes, the petty wars of rival princelings, and the competition for dominance among a host of early city-states. Each of these in time added a unique ingredient to the Russian character and to the story of *Kremlin Gold*.

Between the sixth and eighth centuries, tribal groups of East Slavs from Central Europe migrated to the Russian steppe and settled along the Dnieper River. These were pre-Christian peoples united by an Eastern Slavonic language with Indo-European roots. In the year 855, however, Christianity made its first inroads when two missionaries from Byzantium, Cyril and Methodius, begin translating Biblical texts and church services into the Slavic tongue. They also developed the Cyrillic alphabet and a literary language, Church-Slavonic.

The native Slavic element met a powerful counter-force when Viking expansion in the ninth century brought Nordic adventurers south into Slavic lands. In 860 the Viking chieftains Askold and Dir raised an army of Vikings and Slavs and led it down the Dnieper River to the Black Sea. Two hundred ships sailed into the Bosporus to attack Constantinople, capital of the Byzantine Empire, and begin a cultural and political exchange that would have a profound impact on the development of Russia. Subsequently, in 862, the people of Novgorod called in the Varangian (Scandinavian) prince, Rurik, to rule them. Rurik became the founder of the first line of Russian royalty, which lasted more than 700 years.

The first Russian state, however, grew around the city of Kiev, with Oleg, from the Rurik dynasty, as its prince. By the 880s many autonomous cities were loosely united to form the nucleus of the future Russian Empire, Kievan *Rus*, and by 911 a treaty with Constantinople guaranteed safe passage and trading privileges to Kievan merchants dealing in fur, honey, and wax. The treaty assured an extended peace between Byzantium and the emerging Russian state, as the third great cultural thread began to works its way into Russian history.

By 955 Kievan Rus was ruled by Svyatoslav, the first Russian prince with a Slavonic name. His mother, Olga, who ruled as regent until Svyatoslav was old enough to ascend the throne, strengthened the ties between Kiev and Byzantium when she was baptized in Constantinople as one of Russia's earliest Christians. It was Olga's grandson Vladimir, however, who began the wholesale conversion of his country to the new religion. In 988 Vladimir was baptized and married the Byzantine princess Anna, and Eastern Orthodox Christianity became the official religion of Russia. Images of pagan gods were destroyed or "drowned" in the Dnieper River and the citizens of Kiev underwent mass baptism. Vladimir was later canonized.

The power, the authority, and the fervent Christian faith of Kievan Rus were further strengthened in 1019 when Yaroslav, one of Vladimir's sons, became the monarch of Kiev. In 1025 the city of Yaroslavl was founded to honor this prince. In Kiev itself, the Cathedral of St. Sophia was built to mark Yaroslav's patronage of the Orthodox Church.

During the prosperous years of his reign, numerous churches were built in the Byzantine style, many of them, like St. Sophia, modeled on the famous original St. Sophia in Constantinople. At the time of Yaroslav's death, the Russian state was the largest in Europe and his policies had earned him the title "Yaroslav the Wise."

Old Russia found its lasting ideal of the enlightened ruler, however, in Yaroslav's grandson, Vladimir Monomakh of Kiev, who reigned from 1113–25. A man of unique vision in a violent age of internecine wars and fratricidal feuds, Vladimir united the petty princes of Kievan Rus against a greater threat: the tides of incoming nomadic invaders. Vladimir also left a *Testament* to help his sons follow his example, urging them to "Never judge a man by his wealth. Be generous to the poor. Protect widows and orphans." Vladimir's crown became a royal relic: recreated and worn by Russian rulers for the next 600 years, it symbolized benevolent rule and the burden of government borne by the wise prince.

Against this backdrop, the small provincial trading center of Moscow was founded, grew, and began to prosper. In time the state of Muscovy would dominate Russian affairs, but for many years after Kievan Rus waned in the twelfth century the cities of Novgorod, Pereslavl, Vladimir, Chernigov, Rostov and others vied for power. Each left a legacy of beautiful artworks and memorable rulers that make up the early chapters in the story told by *Kremlin Gold.*

1147: First mention of Moscow in Russian chronicles. Yuri Dolgoruky, prince of Suzdal, Vladimir Monomakh's youngest son, fortifies the small provincial settlement of Moscow. **[1]**

1237: Mongol-Tatar domination of Russia, known as the "Tatar Yoke," begins. Batu Khan, grandson of Genghis Khan, leads his army across Russian territory and attacks the city of Ryazan **[2]**, which falls despite strong resistance. After heroic fighting, the cities of Moscow and Vladimir are destroyed and burned to the ground.

1240: Tatars destroy Kiev, Chernigov, Pereslavl and other southern cities. Batu's empire, the Golden Horde, begins its two-hundred year rule over Russia's land and peoples.

[3] Gold *Kolt* Medallion, Ryazan, 12th–13th century. Treasures like this gold medallion were buried in the city of Old Ryazan, razed by the Mongols in 1237.

1276: Daniel, the youngest son of Alexander Nevsky and founder of the Muscovite dynasty, becomes prince of Moscow, then a small and insignificant town. Under his rule Moscow grows into a prosperous city surrounded by thriving farms and attracting many traders. [4]

1296: Moscow recovers from the first wave of Tatar attacks and slowly rebuilds. The city grows around its fortress, or *kremlin*, which marked the earliest settlement of Muscovites. The kremlin is fortified with earthen walls and an oak rampart.

[5] Icon Crown: Bogolubskaya. Moscow, 13th–14th century. This *oklad* once crowned the head of Mary on a golden icon cover, now lost.

1325: Ivan I Kalita ("Moneybags") [6] becomes Prince of Moscow. He accrues great wealth and uses it to expand his territory. Moscow becomes the residence of the Metropolitan of the Russian Orthodox Church – the highest authority in the land at the time.

1328: Ivan I shows his loyalty to the Tatar khans by collecting tribute from other Russian princes on behalf of the Golden Horde. In exchange he gains four decades of peace for his people. Ivan establishes Moscow as the territory's political center.

1359–89: During the reign of Grand Prince Dmitry Donskoy, Moscow grows in strength, influence, and economic stability. Russians withstand several attacks from Lithuania and repulse Tatar raids on Muscovite territory. Dmitry enlarges Moscow's Kremlin and gives it walls and towers of white limestone.

1380: Dmitry Donskoy calls for a holy crusade against the Tatars. The Muscovite army crosses the Don River and overcomes Tatar forces at Kulikovo Field. Dmitry's valor at the Don earns him the nickname Donskoy ("of the Don").

1453: Byzantium falls to the Turks and Moscow becomes the center of Eastern Orthodox heritage. The Russian church breaks free of Greek influence, which leads to a significant degree of political independence for Muscovite Russia.

1462–1505: Reign of Ivan III (the Great), "gatherer of the Russian lands." [7] Ivan creates a centralized state and enlarges Moscow's territory by purchasing or conquering other principalities' cities and lands.

1472: Ivan's marriage to princess Sophia Paleologue, niece of the last Byzantine emperor, completes the transfer of Greek Orthodox authority and tradition to Russia. Moscow becomes the center of Orthodox Christianity.

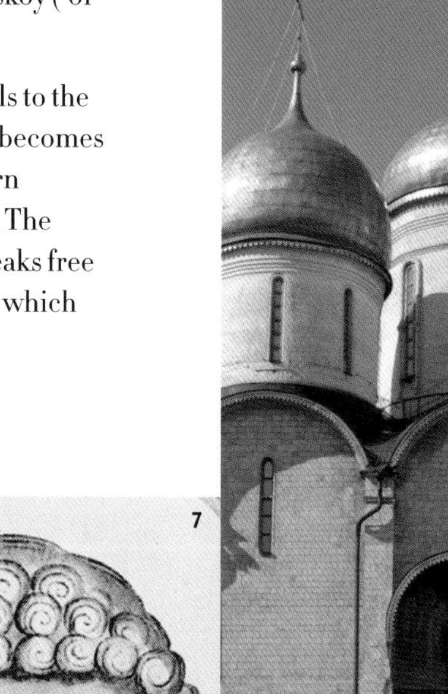

1475

1479: The Cathedral of the Assumption [8] is completed in the center of Moscow's Kremlin. Clad in white limestone and featuring five golden domes, it combines Byzantine and Italian Renaissance architectural styles. It becomes the coronation church for the royal family.

1480: Ivan III leads his army against Khan Akhmed to drive him from Muscovite territory. Ivan refuses to continue paying tribute to the Golden Horde and Tatar domination ends.

1485–1516: New brick walls and towers are built around the Kremlin, expanding it to its present size.

1489: The Cathedral of the Annunciation [9], a private church for the royal family, is built in the Kremlin, next to the Cathedral of the Assumption.

1500

1505–8: The Cathedral of the Archangel is rebuilt. In it are buried the princes of Moscow and the tsars of Russia, with the exception of Boris Godunov, until the founding of St. Petersburg in the early eighteenth century.

1547: Ivan IV (the Terrible) is crowned with the title "Tsar of All Russia." [10] He conducts military campaigns against the Tatars and begins annexation of Tatar lands southeast of Moscow.

1549: The *Zemsky Sobor*, the first national assembly of delegates from around the country, is called by Ivan IV to discuss a program of reforms in a democratic manner.

1550

1550: Ivan IV issues the *Sudebnik*, a code of law in which criminal acts and punishments for each are clearly defined. Citizens' representatives join the judges in executing the laws.

1552: Ivan IV attacks and conquers the Kazan khanate, a powerful remnant of the Golden Horde. The Volga River region is incorporated into Russian territory.

1555–60: To commemorate Ivan's defeat of the Tatars of Kazan and Astrakan, the Cathedral of the Intercession of the Virgin is built at the southern end of Red Square in front of Moscow's Kremlin. Later known as the Cathedral of St. Basil the Blessed, the church features ten domes, each with a different color and design. [11]

1565: Ivan IV creates a royal domain, the *Oprichnina*, in which his power becomes absolute. Russian noblemen are deprived of their estates, which are given to *oprichniki*, secret police officials who dress in black and ride black horses to hunt down traitors. Great landowners and princely families are persecuted.

1581: The southern Cossacks of the Don river region vote Yermak Timofeyevich their senior officer. Yermak captures the capital of the Siberian khanate, thus beginning Russia's expansion into Siberia.

1584: Ivan the Terrible dies and the Russian Orthodox Church begins to regain prominence. In 1589 Job is named the first Russian patriarch, becoming the church's highest authority. Ivan's son Fedor, more devoted to the church than to the task of governing, reigns as tsar under the control of his wife's brother, regent Boris Godunov.

1591: Tsarevich Dmitry, Ivan the Terrible's youngest son, is found stabbed to death in his mother's home outside of Moscow. Rumor claims that Boris Godunov ordered the murder. The martyred Dmitry is later named a saint.

[12] Coffin Effigy of St. Dmitry, Moscow, 1630
This gold effigy covered the shrine of the young St. Dmitry, murdered during the reign of Boris Godunov.

1598: Rurik's dynasty ends with Tsar Fedor's death. The crown passes to his widow, Irina. When she refuses to accept it, Boris Godunov is elected tsar by the *Zemsky Sobor*. **[13]**

1605: Following rumors that Tsarevich Dmitry survived in exile and one day would claim the throne of Russia, "False Dmitry" leads an army of Poles and Cossacks to Moscow. Boris Godunov dies and False Dmitry becomes tsar – one in a series of tsars and pretenders during Russia's "Time of Troubles."

1610: A Polish garrison invades Moscow and settles in the Kremlin. Wladyslaw, son of the Polish king, is elected tsar. The *boyars* (nobles) swear allegiance to him in hopes that he will convert to Orthodoxy and protect the rights of Russian nobles. Instead, Poland tries to rule Russia as a Polish province.

1612: The "Time of Troubles" draws to a close when Prince Dmitry Pozhaysk , a former Russian general, and Kuzma Minin, a wealthy merchant from Nizny Novgorod, raise an army to march on Moscow and expel the Polish invaders. **[14]**

[15] Gold *Kovsch* of Mikhail Romanov, Moscow, 1618. Once owned by the first Romanov tsar, this ceremonial dipper echoes the form of ancient wooden drinking vessels.

1613: Pursuing national unity, the *Zemsky Sobor*, or national assembly, chooses a new tsar – 16 year old Mikhail Romanov **[16]**. Mikhail's father becomes patriarch of the Russian Church under the name Philaret. The Romanov dynasty will last for 300 years.

1645: Mikhail dies and his son, Alexei (the Quiet) **[17]**, becomes tsar. His reign is one of relative stability. He opens new schools and welcomes Western music, art, and architecture to Russia.

1649: Because he must rely on great landlords for defense and support, Tsar Alexei issues a new code of law, the *Ulozhenie*, that legally attaches peasants to their owners, making serfdom hereditary.

1652: Nikon **[18]** becomes patriarch at Alexei's request and introduces Greek practices into the Russian church. He revises ancient Church-Slavonic religious texts. He follows the Greek tradition of making the sign of the cross by putting together a thumb and two fingers instead of just two fingers, in the Russian manner.

1658: Nikon's reforms create a broad opposition of "Old Believers." When he tries to revive Church authority over the tsar, his relationship with Tsar Alexei collapses, and he resigns his office and leaves Moscow. The Church Council later denounces and banishes Nikon, insisting that "to obey the tsar in all political questions" is the duty of the church.

1681: Despite Nikon's disgrace, his reforms endure. Old Believers who continue to cross themselves with two fingers and keep other old rituals are persecuted by the state and put to death. Their leader, Archpriest Avvakum, is burnt at the stake in 1682.

[19] Gold *Kadila*, Moscow, 1616. Tsar Mikhail Romanov presented this gold censer to the powerful Trinity Monastery near Moscow. Its form resembles Russian Orthodox church architecture.

1670: Cossack Stenka Razin leads the Don Cossacks in an uprising against the nobility until he is captured and executed on Moscow's Red Square. Razin becomes a folk hero, his personality and deeds remembered in legends, songs, and poems.

1676: Reign of Fedor III, Alexei's oldest son. Fedor, a progressive ruler, initiates reforms that strip the nobility of their hereditary right to the highest positions in government and the army. Fedor dies at the age of twenty.

Moscow's Red Square in the 17th century **[20]** was the scene of commerce, social diversion, church processions, and executions. Cossack rebel Stenka Razin met his end here in 1671.

1682: Fedor's young brother Ivan and half-brother Peter share the throne in name, though Russia is actually ruled by their sister Sophia **[21]** as regent. As the first woman of power in the Romanov dynasty, Sophia opens the throne to the dynamic female rulers of the next century. During Sophia's regency Peter and his mother are kept outside Moscow and Peter is allowed in the Kremlin only on state occasions.

1689: Political factions supporting the seventeen-year-old Peter overturn Sophia's regency and force her into Moscow's Novodevichy monastery.

[22] Diamond Crown of Ivan Alexeivich, Moscow, 1682. Worn by the brother of Peter the Great, this crown replicates the ancient, symbolic "Crown of Monomakh."

1696: Ivan V dies and Peter I **[23]** becomes the sole tsar of Russia.

1697–98: Peter undertakes his "Great Embassy," a 15-month journey through Western Europe, and becomes acquainted first-hand with European military and naval technology and other Western ideas.

1700: Peter begins the Great Northern War against Sweden to give Russia access to the Baltic Sea and open a "window on Europe."

1701: As part of Peter's educational reforms, the School of Mathematics and Navigation opens in Moscow. Peter initiates other deep and long-lasting cultural, social, and economic reforms.

1703: Peter captures a key position held by the Swedes where the Neva River joins Lake Ladoga. The fortress he builds there, Schlisselburg, will later become St. Petersburg. Peter founds the first Russian newspaper, to report Russian and international news.

1709: Peter wins a great victory over the army of Swedish King Charles XII near Poltava, far to the south of Moscow. He conquers Estonia and Livonia on the Baltic shore and invades Finland.

1712: The seat of government is transferred to St. Petersburg, the new capital of Russia, where it will remain for two hundred years.

1721: The Great Northern War ends when Sweden signs a treaty with Russia, Denmark, Poland, and Germany. Russia annexes the entire Baltic region. Peter the Great is called the first Russian emperor.

1722: Peterhof **[24]** the tsar's summer residence is built in Petrodvorets, a suburb of St. Petersburg. Complete with golden statues, fountains, and waterfalls, this "Russian Versailles" is a dazzling display of Peter's obsession with Western European art and architecture.

1725: Peter creates a charter for the Academy of Sciences, which opens in St. Petersburg after his death in January 1725. European scientists become the first members of the Academy.

1754: Continuing the architectural renaissance begun by Peter, the Italian architect Rastrelli builds the Winter Palace **[25]** in St. Petersburg, which becomes the tsars' residence, and other public buildings in the Russian baroque style.

1755–57: The University of Moscow, the first in Russia, is founded during the reign (1741–61) of Empress Elizabeth, Peter the Great's daughter, as part of her educational program. The Academy of Arts opens in St. Petersburg.

1762: Tsar Peter III **[26]**, grandson of Peter I, is assassinated. His wife, Catherine **[27]**, the future Empress Catherine the Great, succeeds to the throne.

1769: Russia fights against the Crimean Tatars and conquers Moldavia in the Turkish war.

1772: During a time of internal conflict Poland ratifies an agreement among Russia, Prussia, and Austria to divide Poland, with each to rule roughly one-third of its territory. Ancient Russian lands, Belorussia and parts of Ukraine, populated by Christian Orthodox peasantry, become part of Catherine's empire, as does Livonia on the Baltic Sea.

1773: Cossack Yemelian Pugachev claims he is an heir to the throne, and leads a Cossack rebellion. Pugachev's army captures several cities along the Ural River as serfs join the uprising and murder landowners. Pugachev is captured and publicly executed in Moscow in 1775.

1783: Cavalry officer Gregory Potemkin conquers Crimea for Catherine the Great, opening new southern provinces for foreign settlers from Poland and newly acquired Baltic lands.

1785–86: The Charter of Nobility is issued, giving nobles absolute power over their serfs. Catherine, attempting to develop education beyond the traditional church schools, also issues the Statute of Popular Schools, which establishes primary and secondary secular schools.

1796: Paul I **[28]**, son of Catherine the Great and Peter III, becomes tsar. He bans the Charter of Nobility, declaring Sunday a day off for serfs and letting them work three days a week for their owner and three days for themselves.

[29] Gold and Silver Salver, Moscow, 1780.
The allegorical decorations on this presentation plate celebrate the reign of Catherine the Great.

1801: When Paul I is murdered by court conspirators, his son, Alexander I, becomes tsar. Raised and educated by Catherine the Great, Alexander shares her political goals and reinstates the Charter of Nobility, although he accompanies it with a law allowing peasants to buy freedom and land from their owners.

1802: A Ministry of Education is formed with the goal of educating the peasantry. The middle and upper classes strive for higher education. University life flourishes.

1812: Napoleon **[30]** invades Russia and enters Moscow. Largely deserted by its population, the city is set on fire at the Moscow governor's order. Cheated of conquest, Napoleon's army is caught without food or supplies and forced to retreat. Many thousands die on the long winter march to France.

1814: Joining the Prussians and Austrians to resist Napoleon, Tsar Alexander I **[31]** enters Paris in triumph.

1825: When the childless Alexander I dies with no heir, liberal Army officers launch the ill-fated Decembrist uprising, intending to create a constitutional monarchy. After the uprising is suppressed, Alexander's younger brother Nicholas I becomes tsar. Calling serfdom "an evident evil," Nicholas I lays the foundation for the liberation of Russian peasants.

[32] Gold and Gemstone Snuffbox, St. Petersburg, c. 1800. St. Petersburg's fashionable socialites traded luxurious keepsakes like this snuffbox.

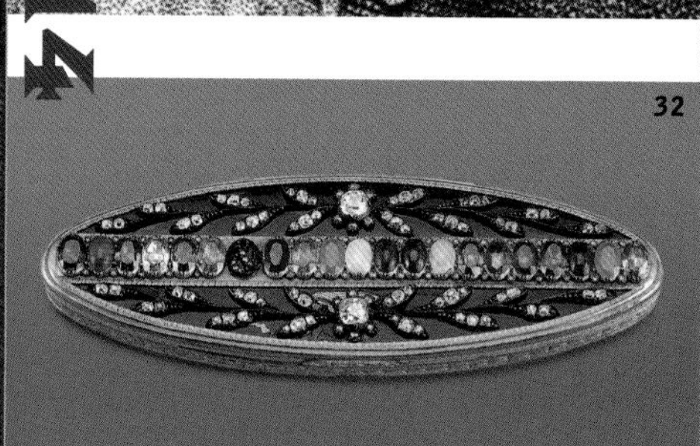

1830s: Poet Alexander Pushkin ushers in a Golden Age of Russian literature. Among the great writers who follow him are Nikolai Gogol, Ivan Turgenev, and Mikhail Lermontov.

1836: Composer Mikhail Glinka begins the grand tradition of Russian opera with his works *The Life of a Tsar* and *Ruslan and Liudmila*. Russian classical music begins to flourish.

1849: Architect Konstantin Thon completes the reconstruction of the Moscow Kremlin begun in 1813 after the great fire. Integrating the Kremlin's ancient palaces and churches into a unified architectural ensemble, Thon creates the Grand Kremlin Palace **[33]**, the Arsenal **[34]**, and the Armory Ward in an elegant Russian classical style.

1853–56: The Crimean War erupts when Turkey refuses the demand of Nicholas I for Russian protection of Orthodox believers on Turkish soil. French and British naval forces enter the Black Sea to fight on the Turkish side. Despite the heroism of its troops, the Crimean War is a disaster for Russia.

1855: Upon the death of Nicholas I, his son Alexander II **[35]** ascends the throne and immediately signs a peace treaty to end the war. Russia loses southern territories previously annexed from Turkey and gives up the right to keep naval bases in the Black Sea.

1861: Alexander II (the Liberator) signs the Emancipation Statute, liberating the serfs.

1863–69: A time of giants in Russian literature. Leo Tolstoy writes his epic historical novel, *War and Peace*, describing Russia's war with Napoleon.

1866: Fedor Dostoyevsky's *Crime and Punishment* is a breakthrough in the genre of the psychological novel.

1881: A terrorist bomb kills Alexander II and brings about the reactionary policies of his son and successor, Alexander III. Ethnic and religious minorities throughout the country are deprived of their rights and most progressive reforms are subverted.

1891: Construction begins on the Trans-Siberian Railway **[36]**, which runs through the vast lands of Siberia up to the Pacific Ocean and connects European Russia with the Far East.

1894: Alexander III dies and his son, Nicholas II **[37]**, becomes the last tsar of Russia.

1898: Private theaters emerge in Moscow and St. Petersburg. At the Moscow Arts Theater, directors Stanislavsky and Nemirovich-Danchenko stage the first performance of Anton Chekhov's "The Seagull." The Stanislavsky school becomes the leading theater tradition in Russia.

1904–5: The Russo-Japanese War pits Russia against Japan as rivals for control of Manchuria and Korea. The war exhausts both sides, buts ends in defeat for Russia.

1905: The first Russian Revolution. On January 9, a large, peaceful demonstration calling for civil liberties is shut down by the police. Hundreds are killed, Nicholas II is called "Bloody Nicholas" by his people, and the day becomes known as "Bloody Sunday". A wave of strikes follows. The workers of St. Petersburg set up a council (soviet) to coordinate the movement for the working-class. Soviets are created in Moscow and other cities.

1905: A Siberian peasant, Gregory Rasputin **[38]**, passes for a holy man in St. Petersburg and attracts the attention of Tsarina Alexandra, Nicholas II's wife. Persuaded that Rasputin can treat her only son, Alexei, for hemophilia, Alexandra admits him to the court, where his scandalous behavior, debaucheries and violent nature cause deep rifts. Some high officials are removed at his word, while his relatives and friends are elevated to important positions. In 1916 Rasputin is murdered by court conspirators.

[39] Fabergé Easter Egg with the Royal Yacht "Standart", St. Petersburg, 1909 (detail). Carl Fabergé's firm created fifty-three extraordinary Easter eggs for Tsars Alexander III and Nicholas II, as presents for their wives.

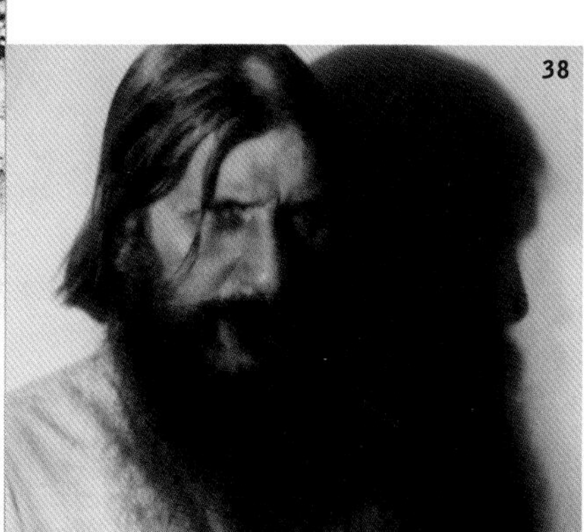

1914: Russia enters World War I after Germany declares war. The German-sounding name of St. Petersburg is changed to Petrograd.

1917: In February, striking workers precipitate a revolution to overthrow the tsar. A provisional government rules the country. Tsar Nicholas II abdicates and the Bolshevik party of Vladimir Lenin **[40]** and Leon Trotsky seizes power in Petrograd in the name of the Soviets.

1918: Lenin moves his government to Moscow, which once more becomes the capital of Russia. The new Russian government signs the treaty of Brest-Litovsk with Germany, losing Finland, Estonia, Latvia, Lithuania, Poland, Ukraine and part of Belorussia. Tsar Nicholas and his entire family are executed by the Bolsheviks in July.

1918–21: Civil War breaks out between the Bolsheviks, known as the Reds, and the monarchists and their allies, known as the Whites.

1921: After the Red Army defeats the Whites, workers' strikes and peasant revolts shake the new regime. Lenin's New Economic Policy grants a limited free market system to peasants, though the new measures cannot avert a severe famine in Central Russia that kills five million people. Church property is confiscated and religion is officially prohibited outside of the home.

1922: The Union of Soviet Socialist Republics is formed.

Moscow in the 1920s **[41]** was the capital of the new Communist state and the center of a growing military and industrial power.

1928: Joseph Stalin **[42]** comes to power, ends the New Economic Policy, and substitutes his own Five-Year Plan and policies of agricultural collectivization. Peasants lose their property and thousands of Russians are exiled to Siberia. An oppressive security regime and sweeping political purges become the hallmark of Stalin's reign.

1934: Sergei Kirov, a popular party leader in Leningrad, is assassinated. The assassin, Nikolaev, is sentenced to death. Large-scale investigations of counter-revolutionaries, arrests and executions follow.

1935: First Moscow subway line opens, featuring brilliant lights, opulent decorations, marble, rhodonite mosaics, and statuary. **[43]**

1936–38: The Great Terror begins with purges in the highest ranks of the Red Army and trials of party opposition leaders. Oppression of the intelligentsia.

1941–45: The Soviet Union participates in World War II. **[44]** Germany invades Soviet territories without a declaration of war. Leningrad withstands a German siege at the price of many thousands of lives. Moscow is never taken by the German army.

1953: Death of Joseph Stalin.

1955–1964: Nikita Khrushchev **[45]** becomes the leader of the Communist party, abandons Stalin's policies, and condemns Stalin's legacy of persecution and oppression. The "Khrushchev Thaw" begins in 1956.

1957: The space age begins in earnest when Russia launches the world's first satellite, Sputnik. **[46]**

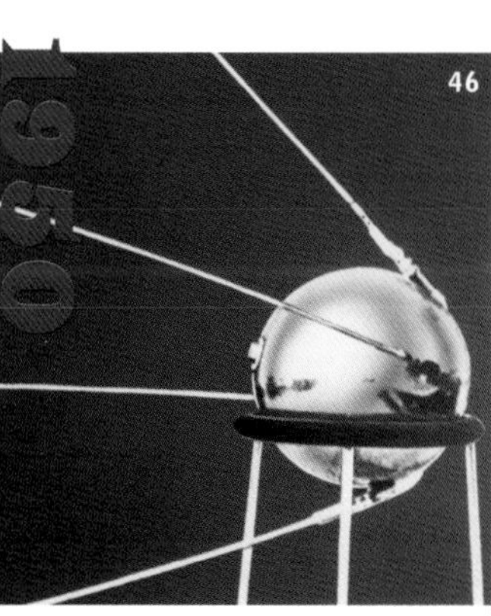

1950

1961: Russian astronaut Yuri Gagarin **[47]** is the first man to orbit the Earth.

The Palace of Congresses is built in the Kremlin. During construction, archaeological remains of the ancient Kremlin fortress are unearthed.

1964: Khrushchev is ousted from power. His successor, Leonid Brezhnev **[48]**, leads the Communist party and the Soviet nation for nearly two decades.

1960

1985: Mikhail Gorbachev becomes first President of Russia and begins profound reforms in the Communist party and society. He proclaims *perestroika*, a reconstruction of the existing regime, and *glasnost*, freedom of speech.

1980

1991: Boris Yeltsin puts down a coup against Gorbachev and suspends the Communist Party. Several Soviet republics declare independence. The Soviet Union collapses. Yeltsin is elected President of Russia.

2000: Russia struggles to overcome economic hardships while pursuing democratic reform. **[49]** Yeltsin resigns and Vladimir Putin becomes acting president.

[50] Diamond and Amethyst Crown, Moscow, 1990. This presentation crown echoes ancient Russian motifs while looking to the future.

2000

THE KREMLIN UNEARTHED: BEARING WITNESS TO RUSSIA'S PAST

N. S. Vladimirskaya,
Deputy Director of Scientific Research
State Historical-Cultural
Museum Preserve, Moscow Kremlin

For more than 500 years, the historical nucleus of Moscow, its Kremlin, has been the focal point of Russian statehood and spiritual culture. The country's oldest museum (officially known as the "State Historical-Cultural Museum Preserve" since the 1980s), the Moscow Kremlin has the supreme honor of preserving and augmenting, as well as studying and popularizing, the unique treasures of Russia accumulated over the centuries. The museum's collections are a material expression of the historical memory of the Russian people and a genuine encyclopedia of the arts.

Today, the Moscow Kremlin is known throughout the world as a significant historical and architectural ensemble. It houses priceless collections of state regalia, the unique Kremlin arsenal, centuries-old Russian paintings, and masterpieces of the finest gold- and silverworks from ancient Rus, Western Europe, and Asia. The thematic and chronological ranges of the Kremlin's historical and artistic collections are uncommonly broad.

The twelfth century marked the appearance, in the center of northeastern Rus, of the small wooden city-fortress that two centuries later would be transformed into the capital of a minor principality and subsequently into the capital of a centralized Russian state, the mighty fortress-stronghold, the Kremlin of Moscow. The earliest information we have about the ancient history of Moscow comes from archaeological discoveries made during excavations of Borovitsky Hill. Relics of long-ago epochs uncovered there – jewelry, coins, arms, and household objects – now are some of the most valuable pieces in the Kremlin's collection.

Treasures found in those hidden troves are priceless, for they document Russian history of the twelfth and thirteenth centuries, a period otherwise described only in brief entries in Russian chronicles. It would be difficult to find more eloquent witnesses to the rare talents of Russian goldsmiths, to the originality of ancient Moscow's material culture, and to the multifaceted education of the city's first inhabitants than these objects.

A major archaeological event occurred in the spring of 1988 with the unearthing of the Great Kremlin Trove, treasures that had lain in the earth for 750 years. They belonged to a prince who had witnessed the tragic invasion of Moscow by Mongol-Tatar hordes in 1238. The 300 works in this hoard made by unknown jewelers of antiquity set a standard for judging the level of ancient Russian art. The treasure's owners were well acquainted with jewelry crafts of the Far East, of southern Russian regions, and of the Baltic Sea region. The face of a gold ring found in the hoard is adorned by an Arabic inscription that reads "Glory and success, and power, and happiness, and adornment to the bearer of this." These words, by an anonymous master, might well stand as an epigram for the opening of this exhibition and serve to inspire its visitors.

As we stand on the threshold between millennia, our understanding of the contribution of Russian art to the artistic heritage of the world takes on new significance, and dialogue between countries and peoples becomes ever more vitally indispensable. To this the present exhibition bears indisputable witness.

ESSAYS

RUSSIAN JEWELRY ARTS OF THE TWELFTH TO THE SEVENTEENTH CENTURIES

M. V. Martynova

From the distant past to the present, people have been drawn to the beauty of brightly shining gold and softly glowing silver, captivated by the bright tones and magical play of light on precious stones and by the shimmering opalescence of pearls. They have attempted to extract from nature elements of her wealth – to adorn themselves, their dwellings, and their places of worship and to delight in nature's splendor.

The art of jewelry occupies a significant place in the artistic traditions of many cultures. In every epoch craftspeople have infused their works with their artistic ideals and their ideas of beauty, continually perfecting techniques for processing precious metals and gems. They learned to create embossed relief and sculptural images, to draw the finest wire and to fashion from it patterns as light as lace, and to engrave on metal with a sharp chisel. They invented enamel and niello – glassy and metallic alloys with which they colored the surface of gold and silver objects – and learned to embellish their works with precious stones. As in so many cultures with ancient roots and traditions, the jeweler's art is an integral part of Russia's artistic heritage.

The oldest cultural centers in Russia's European regions were the cities of the northern Black Sea coast, which were settled by Greeks as far back as the middle of the first millennium B.C. High-quality jewelry and other works of art and crafts were produced there. Masters working in the Greek colonies executed commissions for the chiefs of nomadic tribes and peoples who, one after another, populated the broad steppe expanses of the

Dnieper River area. Over the years, the art of these jewelers underwent a profound transformation, while at the same time preserving the traditions and techniques of its ancient heritage. Illustrating this transformation are the objects found in a treasure trove discovered in 1927, apparently made for members of the tribal aristocracy of the Huns who had encroached into East Europe from the Ural region in the fourth century A.D. This treasure, found by chance in the Kursk region at the upper reaches of the Suzdhal River and characterized by heavy gold jewelry ornamented with colored pieces of glass, is made in the so-called polychrome style [*plates 1, 2*]. This art played its part in the emergence of the proto-Slavic culture that colored the earliest phases of Russia's history.

These heirlooms of a restless epoch that witnessed the continuous migration of peoples across Eurasia also mark the early stages in the formation of Byzantine art. The Byzantine empire ultimately would contribute richly to the world's cultural heritage. It had an enormous influence on the artistic development of many countries during the Middle Ages, including ancient Rus, which, after adopting Christianity in 988, actively began to appropriate and creatively transform Byzantine artistic traditions.

Byzantine artifacts preserved in Russian museums bear witness to the close political and cultural ties that linked the Byzantine empire and the young Russian state. These contacts were particularly active from the tenth to the twelfth centuries, when Byzantine art was at its peak. It was then that numerous examples of Byzantine crafts found their way into Rus, among them renowned partitioned enamels [*plate 10*] and splendid examples of glyptics (carving on stone). Two cameos in the Armory collections, exceptional for their size and artistic quality, date from this period. One of them, "Christ Giving His Blessing," executed in a type of jasper known as bloodstone and mounted in a gold frame, is included in this exhibition [*plate 4*]. In addition, a tomb erected in antiquity on the burial site of the Christian saint Dmitry of Thessalonika [*plate 3*], which did not survive to the present, is represented in the form of a small reliquary that demonstrates the legacy of Byzantium's silversmiths.

The eleventh to thirteenth centuries in Rus were marked by an increase in the arts and crafts, prominent among them the jeweler's art. Archaeologists' discovery of hidden treasures has provided the most important material for the study of that art. Though often buried during years of internal strife among warring princes or during raids by nomadic tribes, the majority of these treasures were hidden during the period of the Mongol-Tatar invasions of Russian lands.

The hidden troves of this era are composed almost entirely of women's jewelry; most of these items date from the twelfth and early thirteenth centuries. These priceless artifacts testify to the high level of artistic culture of that time, to the originality and maturity of aesthetic thought, and to the free use of the most varied technical approaches.

Even at that early date, Russian jewelers knew the secrets of embossing, niello, the finest filigree, and of small silver balls soldered to filigree. They had mastered one of the most complicated jeweler's techniques, partitioned enameling. Images executed with this technique adorn the unique objects dating from pre-Mongol Rus in the Ryazan treasure found in 1822 at the site of Old Ryazan, a city entirely destroyed by the Mongol-Tatars in 1237 [*plates 11, 12*]. The items in this trove – massive necklaces accompanied by pendants and medallions of many shapes and sizes – have no close parallel in the jewelry art of the pre-Mongol era. Judging by the richness of the materials and the masterful execution, the items belonged to members of a princely family. In addition to the early appearance of the images of saints on these works, one also is struck by the variety and beauty of the filigree patterns that decorate the Ryazan jewelry. In some instances the filigreed patterns rest calmly on the surface of the metal. More commonly they rise in multi-tiered, lacy layers on a golden background decorated with rounded precious stones, which shine brilliantly in elegant raised settings. In this technique of stone mounting and in the design of the filigree patterns a certain similarity to examples of Roman art may be observed. This attests to the active artistic contacts that existed between Rus and Western European cultures.

The flourishing of Russian art during the twelfth and early thirteenth centuries was interrupted by the Mongol-Tatar invasions, a time of tragic ordeals for the Russian people and their culture. During this time of subjugation, known as the "Tatar Yoke", the Tatars pillaged Russian lands, destroyed cities, reduced countless churches to ruins, and damaged thousands of precious objects. Russia's native artistic traditions, however, withstood Tatar incursions, with cultural life continuing in some principalities, trading quarters, and monasteries.

A special place in the history of Russian art belongs to Novgorod, whose location spared it the depredations of foreign invasion during this period. Archaeologists have found evidence of leather tanning, bone-carving, jewelry-making, and woodcarving workshops in this city that, as early as the tenth century, had become an important center for trade and crafts. Among examples of the applied arts from Novgorod are many small works: miniature icons, images, crosses, and panagias carved from wood, bone, and stone or cast in silver and copper [*plate 16*]. These items were produced in great quantities not only for churches and monasteries but, first and foremost, for private individuals. Some of them, attaining the level of high art and meticulously cared for, were passed down as family heirlooms from one generation to the next.

Toward the end of the fourteenth century, Moscow assumed the role of unifier of the Russian lands, spearheading the struggle of the Russian people against the domination of foreign powers. With Moscow's political ascent began the emergence of its arts. It is significant that the arts in Muscovy at that time drew heavily on the traditions of the earlier era of independence that preceded the Mongol invasions. In all areas of culture, heightened interest in pre-Mongol Rus and its artistic heritage is evident; jewelry-making of the early fifteenth century was no exception. Thus, it was not by chance that, when creating a setting for the "Madonna of Bogolubskaya" icon of the Moscow Kremlin's Church of the Annunciation, the jeweler included within the design a crown created during the pre-Mongol era [*plate 15*]. Its filigree ornament is similar in pattern and technique to the filigree work found on pieces in the Ryazan treasure trove. The fourteenth century Muscovite master who restored the ornate crown added precious stones mounted in settings typical of that earlier time.

The formation of a centralized state, essentially complete by the end of the fifteenth century under Ivan the Great, was marked by the widespread construction of stone buildings in Moscow. It was at this time that the ancient wooden towers and palisades, so often built and rebuilt after devastating fires, were replaced by the majestic stone ensemble of the Moscow Kremlin. The new white limestone walls and palaces, which housed Russia's rising class of nobles as well as the ruling grand prince and his court, stood as compelling symbols of the authority of the state. As the power vested in the Moscow Kremlin grew, the majesty and authority of the church grew along with it. For solemn services in the newly built cathedrals in the heart of the Kremlin, a splendid array of church utensils was created. The reliquary known as the "Small Zion," for example, commissioned by Ivan III for Russia's principal cathedral, the *Uspenskii Sobor*, or Assumption Cathedral, echoes the form of Moscow's early architecture [*plate 17*]. On its frontal facet is a triple icon: the Savior, the Mother of God, and John the Baptist; on the three other facets appear figures of the apostles.

In the sixteenth century, Moscow, by now the political and economic center of the country, also became the pan-Russian center of culture, a status it would long retain. In the workshops of the Moscow Kremlin, a host of artists and calligraphers, carvers and embroiderers, weapon-makers and metal casters summoned from all parts of Russia combined artistic skills of the highest order. In the tsar's silversmith workshop, craftsmen labored to create precious works of art intended to bestow splendor on majestic and solemn court ceremonies and church services. Those luxurious items were essential in their time for reinforcing the power and authority of the ruling regime, whether that of the early grand princes or of the tsars who followed. Today they are valuable as legacies of a national art that reflects the aesthetic ideals of talented native craftsmen and women. Unfortunately, the names of almost all the master goldsmiths and silversmiths working in Moscow in the sixteenth century remain unknown. Virtually no information about jewelers was preserved in the written sources that have survived to our time, and their works bear no hallmark or author's signature.

In the Kremlin workshops of that period, foreign craftsmen worked alongside Russian silversmiths; their numbers increased particularly during the second half of the century. Collaborations between local goldsmiths and Western European masters of the Renaissance also affected the Russian jeweler's art.

Renaissance influence, visible in the design of many sixteenth century artifacts, is especially evident in ornamentation, which as a rule featured elaborately stylized plant elements. Those compositions developed precise, rhythmical patterns around a clearly defined basic motif. The refined coloration of the enamel, velvety niello, and large precious stones completed the artistic effect of these objects.

Master jewelers in the sixteenth century perfected a number of techniques for working with and decorating objects in gold and silver. One of the more widespread methods of ornamentation was niello, a special black alloy of precise proportions of silver, copper, lead, and sulphur that was used to fill the lines engraved in a metal surface. Russian niello engraving on gold, which flourished during the second half of the sixteenth century, adorned liturgical vessels, settings of icons and Gospels, pendants sewn onto secular and church vestments, and clerical headgear. Niello was employed more frequently by Russian jewelers than by their counterparts in Western Europe.

Niello engraving of this period, during the reign of Ivan the Terrible, is distinguished by exquisitely proportioned figures with beautiful silhouettes and by impeccable technical execution. This technique, featuring deep black tones without the slightest runs or defects, distinguishes the drawing on a magnificent small reliquary commissioned by Tsar Fedor Ivanovich for his wife, Irina [*plate 19*]. The figure of Saint Irina, the tsarina's heavenly guardian, was executed by an unknown master whose hand is evident in numerous outstanding engraved niello objects of the late sixteenth century.

Enamel techniques also experienced a renaissance at this time. Whereas fourteenth- and fifteenth-century enamel colors were of limited range, with enameling playing a secondary decorative role on gold and silver objects, it now blossomed as a primary technique. More and more often enamel began to have a soft, light color scheme dominated by white and light blue hues of many rich shades, from intense cornflower to subtle turquoise. Deep reds were often introduced as accents. Enameled plant ornaments in trefoil patterns were characteristic of Muscovite jewelry of the sixteenth century. Lilies and many-petaled flowers on softly curving stems, laid with gold and silver threads and colored with enamel, were also common. These enamel patterns were usually framed with filigree.

The setting for the icon cover "Our Lady of Odighitria" [*plate 18*], a family icon of Ivan the Terrible from the Moscow Kremlin's Archangel Cathedral, is a splendid example of late sixteenth-century enamel art. Its setting is decorated with a graceful, stunningly beautiful pattern of filigree and enamel. Though the applied ornament is soldered to the background in many places, that of the crown and frame was cut instead from a thin sheet gold, framed in filigree, and sealed in enamel to create a unique effect. Although the range of colors used in this object is almost motley, the work as a whole is surprisingly coherent, the enamel colors resonating with those of the adjacent pearls and precious stones - sapphires, rubies, tourmalines, and emeralds. Small balls of gold are melted into the enamel like frozen glitter splashed onto the ornament.

Glyptics, the art of carving images on stone, was another technique widely used in the sixteenth century. Although we have no precise information to date the origin of ornamental stone carving in Russia, this craft had a long history and was already fairly well developed by the sixteenth century. Catalogues of the tsars' possessions mention Russian-made carved stones, and surviving works testify to the prevalence of the craft. A place of honor among these works is held by the cameo "John the Baptist" [*plate 20*].

The evolution of Russia's arts was interrupted at the beginning of the seventeenth century by the so-called "Time of Troubles", a period of violent political and social discord that saw the invasion of Russia by Polish-Swedish armies and the appearance of numerous pretenders to the tsar's throne. After the expulsion of the invaders in 1612, Russia underwent a gradual rebirth, and the Kremlin art workshops, including the Silver and Gold Chambers, resumed their operations and once again became leading centers of Russian jewelry production.

The wealth of material available from seventeenth century archives offers us a broad look at the activities in the Kremlin jewelry workshops and gives us the names and origins of many artisans. It is known that craftsmen working there came from cities in central Russia and from cities of the Volga region, as well as from northern centers. In the middle of the century, expanding cultural connections with Ukraine and Belarus brought specialized craftspeople from these regions to Moscow. This influx of artisans influenced the design of the crafts produced in Moscow's workshops and, in turn, enriched the newcomers with new creative encounters. Although goldsmiths and silversmiths of the seventeenth century very rarely signed their creations, comparing surviving artifacts with archival documents allows us to determine the authorship of some of the more significant works.

Sixteenth-century traditions are very clearly manifested in Muscovite jewelry art for most of the first half of the seventeenth century. Technical execution, color schemes, and ornamental motifs often resemble creations of the previous century, apparently in deliberate emulation of older masters. Thus, a censer created in 1616 by artisans Danila Osipov and Tretyak Pestrikov [*plate 22*] replicates the form of a censer, not included in this exhibition, made in 1598 on order from Tsarina Irina Godunovna.

The principle of following an illustrious example when creating a work of art is characteristic not only of the jeweler's art but also of ancient Russian art in general. Yet, far from blindly copying works of the past, Russian craftsmen freely transformed them. Thus, while reproducing in the mid-sixteenth century a renowned fifteenth-century *panagia* from Novgorod, a later master added variations to the original. In the decoration of a gold censer of 1616, in contrast to that of its sixteenth-century predecessor, embossing was used rather than niello, to create the impression of an entirely original work of art.

Embossing plays an important role in works of Russian jewelry. The cover of Tsarevich Dmitry's shrine, an extremely rare example of Russian silverwork [*plate 28*], reflects the virtuosity of embossers of the first half of the seventeenth century. Such depictions of the saints made in precious metals had been known in Rus since the second half of the fifteenth century. In the early seventeenth century, the Kremlin workshops produced several silver shrines, among them this one to house the remains of the youngest son of Ivan the Terrible. Documents reveal that, in the eighteenth century, its cover, with a depiction of the young saint, stood vertically alongside a painted likeness of the Tsarevich in the same icon case; in the nineteenth century, like a splendid icon image, it was placed directly over the coffin of Dmitry in the Kremlin's Archangel Cathedral.

In the seventeenth century, new trends, evidence of the general evolution of Russia's arts, appeared in the Moscow jeweler's art. This process was marked by a growing secularization that began to undermine the medieval world view and the preponderance of religious imagery in medieval art. In the jeweler's art, this was expressed in the creation of a new artistic style – cheerful, brightly colored, and fairy-tale festive. This style of execution is seen in both the secular and liturgical items that survive.

This increasingly prominent secularization also led to the appearance of new directions in theme and ornamentation. Plant motifs, a favorite in the Russian decorative arts, began to approximate more closely the forms of living nature. Large, luxuriant flowers with easily identifiable prototypes started to appear in ornamental designs that previously had been highly stylized. Master silversmiths, like painters, strove for greater three-dimensionality in their treatment of forms, attempting to depict movement and to construct architectural and landscape backgrounds with a greater sense of space than before. Alongside themes from the Gospels are scenes from Old Testament stories and depictions of parables, as well as allegories borrowed from book illustrations.

Enamel suited the aesthetic demands of the seventeenth century and was widely used in jewelry settings. Craftsmen continued to use enamel with filigree ornamentation, although, in contrast to sixteenth century practice, the technique was used primarily to ornament works in silver. The range of colors in these objects varies, though green and light blue tones dominate, in combination with white and black. Colorful surfaces are enlivened by colored dots and applied gilded stars, blossoms, and tendrils. Bands of white enamel globules, imitating strings of pearls, appeared as decoration, and, toward the end of the century, filigree ornaments with enamel became increasingly saturated and opulent.

More often, jewelers began to decorate gold objects with multicolored enamel laid over engraving or covering embossed imagery [*plates 24, 25, 26*]. Enamel on relief work was technically very difficult and demanded from jewelers a high level of skill. The jeweler's work of this period was closely linked with that of the embosser. Seventeenth-century works are also distinguished by extraordinary luxuriance and beauty, often created by a rich and varied range of colored enamels.

Popular decorative schemes combined opaque enamels with transparent enamels of intense and lustrous tones: red, deep blue, emerald green, turquoise. Background gold, shining through these enamels, intensified their brightness and depth and made them seem to glow from within.

In the first half of the seventeenth century, at the bidding of Tsar Mikhail Romanov, the first tsar of the new Romanov dynasty, Kremlin workshop masters fashioned regalia, presentation arms, and precious utensils designed to restore the splendor that had graced the Russian court from the fourteenth through the sixteenth centuries. Many of those items were brightened with enamel. Among them are the miniature *bratinka* winebowl that belonged to Tsarina Yevdokia Lukianovna [*plate 24*] and a set of hoods for bows and arrows, decorated with a plant-and-flower pattern and heraldic imagery [*plate 25*].

Included here are many pieces of jewelry linked to Moscow's sovereigns, as attested to by the inscriptions that adorn them in accordance with Russian tradition. An ornate inscription on a miter [*plate 29*], for example, informs us that this headpiece was made for the Metropolitan of Rostov in 1634 by order of Tsar Mikhail. Painting on enamel plays an important role in the decoration of the many pendants adorning the miter. This technique, which was first used in the seventeenth century and which offered great expressive possibilities, gained wide popularity in the second half of the century. On gold items it often appears in combination with other enamel techniques, such as, for example, the ones we find on the setting for the "Trinity" icon [*plate 32*], and on the lovely *bratina* and small plate [*plates 36, 37*] presented to Tsarevich Alexei for his birthday by his grandmother and by his father, Peter the First.

Painted enamel, usually embellishing silver objects, reached its peak of popularity at the end of the seventeenth century. The northern city of Solvychegodsk was the major center for the production of such works. Jewelers there rendered floral shoots and garlands, depictions of people and animals, and even complex multi-figured compositions in multicolored enamel paints on a ground of white enamel [*plates 41, 42*]. The drawing and shading were executed in black or red-brown enamel. Bright, luxurious flowers, frequently tulips, dominate painted enamel works in the Solvychegodsk style; in this one can discern the influence of the floral style popular in Western European Baroque art.

In the ornamental vocabulary of techniques such as niello and carving, which often appear together in the second half of the seventeenth century, a wide range of flowers were also popular. At the same time, Eastern motifs, well known to Russian masters from imported Turkish and Iranian jewelry and fabrics, were occasionally used in the finishing of niello and carved objects. Indeed, on many works of this time Russian artists' links with their counterparts in other countries stand out vividly. For example, a plate in the exhibition is adorned with lush floral garlands in the spirit of the European floral style, while the glasses, also presented here, with bunches of flowers and fruits hanging from ribbons, echo European Baroque art [*plates 46, 47*]. The ornamental motifs are positioned on a finely patterned niello ground, a signature of works by Eastern masters.

A dazzling profusion of precious stones completed the decoration of much old Russian gold- and silverwork. The memoirs of foreigners visiting Rus in the sixteenth and seventeenth centuries describe with rapture and delight the incalculable riches accumulated in imperial and monastic treasure-houses and the splendor of the tsar's regalia, clothing, clerical vestments, and secular and sacred utensils, all covered in priceless materials. Those reports, and the great quantities of precious stones on the surviving objects, are all the more surprising since precious stones were almost never mined in Russia before the eighteenth century but were imported instead from India, Burma, and China. From there to Rus came a variety of stones of various shades of red: rubies crimson as blood, cherry-red garnets, pink tourmalines, deep red spinels. In Ceylon, rare and beautiful stones – ruby, garnet, amethyst, topaz, and the cornflower-colored sapphire greatly esteemed in Rus – were mined as early as the thirteenth century. Blue turquoise was bought from Persia, while emeralds were imported from Egypt and the Arabian peninsula. The fabulously rich Colombian mines were discovered in the sixteenth century. India was the only source for brilliant diamonds.

Traditionally, Russian people in their legends had sung praises to the beauty of precious stones. The popular imagination imbued them with magical properties, giving rise to beliefs in the mysterious power of precious stones over the thoughts, feelings, and fate of humans.

Emeralds were honored as stones of wisdom. The power to reveal deceptions and banish fears was attributed to sapphires. Folk legends intoned "Whoever shall carry with himself sanguine sapphire [*yakhont chervchatyi* – as rubies were called in ancient Rus] shall see neither fearful nor wicked dreams." In the old days it was believed that turquoise and coral lose their luster when laid on the hand of a mortally ill person, that diamonds tame fury and the love of power, and that if a warrior wore diamonds mounted on weapons carried on his left side, he would not be killed.

Russian jewelry reveals everywhere the passionate love of bright, cheerful precious stones. To the jewelry artist, gemstones are not merely symbols of wealth; they are infused with the aura of beauty, radiate light as does no other material, and provide the jeweler with a uniquely expressive medium with which to realize his or her artistic conceptions.

Before the seventeenth century, Russian master jewelers embellished their creations with rounded, uncut stones known as cabochons that preserved the natural form of the gem. In the seventeenth century, however, faceted stones appeared alongside such rounded cabochons – brightly sparkling gemstones that intensified the color saturation of the objects they adorned . At that time, small stones laid in ornamental stripes and studs were used to decorate a wide range of objects. Dazzling examples are the celebration crowns made at the end of the seventeenth century for the young tsars Peter and Ivan Alexeivich, one of which appears in this exhibition. Covered in sparkling diamond studs in the shape of rosettes, with crowns and double-headed eagles, this piece resembles a blinding, flaming hemisphere that seems to radiate a solid field of brilliant light [*plate 35*].

Moscow's jewelers also created a great deal of precious tableware. In the words of the Danish ambassador Jacob Ulfeld, during a 1575 feast at the palace,

On all the tables so many dishes of silver and cups were laid that there was no empty space, but dish lay on dish, glass on glass.

In the early seventeenth century, another foreign guest described the setting at the tsar's formal reception:

Cups and goblets were aglow with rubies, sapphires, hyacinths, corals, and pearls; the entire table blazed with gold and precious stones.

The Armory contains the only collection of its kind of old Russian precious dishes of the sixteenth and seventeenth centuries, whose shapes recall the wooden and ceramic utensils long in use among Russian people of all social strata at the time. In this collection, a special place belongs to the gold and silver ladles [*plates 21, 23*] that were used at feasts to drink mead, a much-loved drink in ancient times. Meads, known in Rus since antiquity and mentioned in chronicles beginning in the tenth century, were brewed according to differing recipes and infused with a variety of berries and fruits. Consequently, they differed from each other not only in taste but also in color. Red meads were drunk from golden ladles, white meads from silver.

Bratina winebowls and toasting cups also have a place in Russia's national tableware [*plates 49, 50*] and were widely used among both royalty and the *boyars* (nobles). During feasts, a bratina (from "brother") filled with *kvas*, or beer, would be passed from guest to guest, creating communal bonds among those seated together at table. Refined ornamentation and decorative beauty also distinguish other kinds of tableware that was reserved for the nobility: goblets, glasses, and cups made by Moscow's jewelers, who refined the forms and softened the contours of ancient traditional vessels in their quest to create a decorative art that would rival its Western European counterparts.

Russian jewelers infused each of these objects with their aesthetic taste and technical mastery. Although these precious works of art were intended primarily for the elite, their creators brought elements of folk art into the shapes and ornamentation of their works. From this springs both their profound originality and their distinctive national character.

RUSSIAN JEWELRY ARTS OF THE EIGHTEENTH TO THE TWENTIETH CENTURIES

I. D. Kostina

ESSAYS

The early eighteenth century in Russia was marked by important political events and reforms brought about by the direct intervention of Tsar Peter I. Among many other things, those transformations affected the way jewelry manufacture was organized. In 1700, Peter abolished the Gold and Silver Chambers of the Moscow Kremlin and obliged craftsmen who produced items in gold and silver to mark their works with personal hallmarks and to observe assay standards established by the government. Later, hallmarks of large cities were introduced, as were hallmarks of assay masters charged with ascertaining that metal corresponded to an established standard of purity.

In 1724, Peter issued a decree addressing the formation of craftsmen's guilds. In every guild elected elders (or aldermen) were to supervise the artistic aspect of the work produced in the shops. Beginning in the mid-eighteenth century, aldermen of the Moscow silversmiths' guild had their own hallmark attesting to the high level of workmanship of an item. The construction of St. Petersburg and the transfer there of the capital in 1712 led to an influx of jewelers, both Russian and foreign. Many of the foreigners made Russia their permanent home. By the second half of the eighteenth century, St. Petersburg had surpassed Moscow as a major center for the production of works in precious metals.

Strong economic connections with Western Europe, primarily Holland and Germany, helped establish the Baroque style in Russian art. As in many European countries, in Russia the Baroque style acquired its own national traits. Even while copying Western European prototypes, artists introduced into them their own understandings of beauty, and used traditional artistic methods and techniques.

One eighteenth-century innovation in Russia was the appearance of portrait miniatures on enamel. The popularity of the genre was fueled by an emphasis that developed at that time, in both philosophy and the arts, on the human being as a unique individual. At first, the primary patrons (and subjects) of these portraits were members of the royal family. Enamels depicting the tsar and his wife, Catherine, often served as presents or awards. Several such badges, bearing portraits of Peter I in gold frames accented with precious stones, have survived to our day. Portrait miniatures were also mounted on snuffboxes. Grigory Musikiisky and Andrei Ovsov, talented artists of the Moscow Kremlin Armory who were both later transferred to St. Petersburg, were the first to pursue mastery of the demanding art of painting on a small enamel plate. In the portrait of Peter I on a gold snuffbox [*plate 54*], executed by Ovsov after the tsar's death, the artist managed to convey not only an external likeness but also characteristics of the great reformer's personality – his will, intelligence, and decisiveness.

A silver chalice on a tray [*plate 55*], executed in the Baroque style, resembles a soup tureen, though it is hard to imagine bouillon ever being poured into it. The purely decorative function of the vessel is confirmed by the fact that it bears no signs of use. The maker of this chalice employed the ancient art of filigree, which was rarely used in the first half of the eighteenth century as it tended to be insufficiently expressive for the demands of the Baroque style. Here the anonymous St. Petersburg silversmith successfully overcame that barrier by using fairly large-gauge wire and by slightly raising the filigree pattern above the gold surface.

During the Baroque period and that of the Rococo that followed it, the favorite technique was embossing, and the best masters were virtuosi. Works of the 1750s offer fine examples of this technique. A gold and silver tabernacle made in 1753 [*plate 56*] is decorated with beautifully incorporated *rocaille* ornament and multi-figured compositions, while a presentation ladle made in 1755 [*plate 58*], resembling a decorative vase, features a striking relief pattern complemented by cast figures of perching eagles.

The elegant, intricate Rococo style was particularly beloved by well-to-do purchasers of precious wares in Moscow and in provincial towns, where the primary customers for such costly items were ordinary townsfolk and merchants risen from peasant roots. Those essentially conservative classes yearned for the old Russian patterns preserved in traditional folk art and followed new fashions unwillingly; the stricter, more restrained designs of the classicism that came to replace Rococo did not fully correspond to their notions of beauty. They remained loyal to Rococo and continued to choose Baroque ornamentation in silver, particularly in the forms of luxurious tulips and bunches of fruit, almost until the end of the eighteenth century.

Items once belonging to a formerly large service [*plate 60*] are stunning examples of provincial art of the second half of the century. This service was made in Tobolsk for the family of the governor of Siberia, Denis Chicherin. That city, founded in 1587 in the northern part of West Siberia, was in the eighteenth century the administrative center of Siberia and also its center of culture. Tobolsk had its own theater, published a magazine, supported schools and libraries, and sustained many craftsmen, including silversmiths. The second half of the eighteenth century was particularly auspicious for the silversmiths of Tobolsk. In Governor Chicherin they found a well-to-do patron who encouraged them to realize their talents. Items in this tea service are notable for their clear geometrical forms decorated in niello. The themes adorning them illustrate popular literary works, in which entertaining dramas far removed from everyday life unfold between elegant ladies and dashing cavaliers.

The high level of skill of jewelers who worked with gold and precious stones is apparent in the few surviving examples of women's jewelry and also in snuffboxes. The fashion of sniffing tobacco came to Russia during the reign of Peter I and by the mid-eighteenth century had become a widespread obsession among townspeople. Well-made snuffboxes were popular as gifts, rewards, and collectibles, and certain craftsmen specialized in making them. The most famous snuffbox in the Armory collection was commissioned by the Empress Elizabeth, daughter of Peter I, for Count Aleksei Razumovsky, the husband she secretly wed and, by necessity, publicly ignored [*plate 59*].

Looking at this snuffbox, the viewer is surprised by the profusion of diamonds decorating it but even more by the consummate skill evident in its construction. The arrangement of its multi-figured composition is deft, the modeling of the relief confident, the selection and setting of stones meticulous. Another snuffbox, also executed in the Rococo style, is striking for its beautiful juxtaposition of green stone with polished and matte gold, in the midst of which sparkle precious stones. Dark emeralds resonate with the green of heliotrope, in which red inclusions resemble lustrous rubies, and black tones contrast with the diamonds.

The Neo-Classicism that came to replace Rococo led to objects of more severe form and restrained decoration. Snuffboxes in this style display clear geometric forms: the oval, the circle, and the rectangle. Transparent enamel laid on a background engraved with a fine pattern, which shines through the enamel, became a fashionable adornment for gold objects, as demonstrated by an almond-shaped snuffbox from the early nineteenth century [*plate 66*]. The design of the elegant snuffbox by master J.-P. Adore [*plate 63*] is even more subdued: Its lid is adorned only with a diamond monogram of Elizabeth, but, on opening it, one may marvel at a watercolor portrait of the Empress.

Classical restraint was characteristic of St. Petersburg's craftsmen. Works in the classical style by Muscovites, however, are saturated with ornamentation, as if those artists feared the presence of a plain, unadorned surface. The surface of the dish made by prominent silversmith Alexei Ratkov [*plate 61*], for example, is densely covered with the garlands, crowns, and allegorical depictions typical of the neo-classical style, but the shape of the dish, as well as the sinuous contours of the ornamental layers is closer to *rocaille*.

Such juxtapositions were common in the work of Moscow's silversmiths. The best of them could cleverly combine diverse styles and techniques. Ratkov, in his finishing of a silver chalice [*plate 62*], demonstrated skills in embossing, casting, and niello, and further embellished the chalice with medallions containing paintings on enamel. Silversmiths generally did not master the art of painting on enamel, baked over a metal (usually copper) base. The technique required specialized training as well as painting ability; enamel medallions usually were purchased ready-made or commissioned from artists.

A heightened interest in niello technique can be seen in Russian silversmithing of the second half of the eighteenth century. The expressive potential of niello, achieved by the contrast between the gilded silver and the velvety dark tone of niello, attracted both customers and artists. A setting for a Gospel [*plate 64*] with uniquely large niello plates is one of the masterpieces of late-eighteenth century Moscow silversmithing. The artist has skillfully combined a variety of subjects on each cover of the book to create a single, unified composition. Oval medallions enclosed in laurel wreaths are joined by garlands, bouquets, and flower baskets on hanging ribbons, which in turn are linked to decorative urns and bows. The designer of this setting drew on all the most characteristic elements of classical decoration. The background between the niello images, an especially noteworthy technical display, is filled with engraved patterns dominated by the fish scales popular among Moscow's craftsmen. Judging by the hallmarks, two craftsmen executed this setting, intrepid engravers who were not afraid to risk losing their time and labor to the possibility of an unsuccessful firing.

At that time and later, in the nineteenth century, many craftsmen preferred to use the technique of smooth niello engraving, choosing not to decorate the background around the pictures as described above. This was prompted first and foremost by the desire to reduce the cost of the works, but it may also have been dictated by artistic considerations. An example of this practice is a silver pot [*plate 68*] made by the talented craftsman Sakerdon Skripitsyn from the small town of Vologda in the northern European part of Russia. He drew splendidly and, taking print engravings as models, could expertly compose drawings on metal surfaces. On this pot he portrayed a view of the palace square in St. Petersburg, an example of the popularity of cityscapes in the nineteenth century. A surge of nationalistic sentiments dating from the Patriotic War of 1812 and Russia's victory over Napoleon fueled this interest and led to the production of expensive souvenirs and mementos. Muscovite craftsmen naturally featured views of Moscow in decorations. Such products from the firms of Ivan Khlebnikov, Pavel Ovchinnikov, and Vasili Semyonov became quite famous and illustrate the influence these firms exercised on Russian jewelry arts during the second half of the nineteenth and the early twentieth century.

Craftsmen in the Khlebnikov and Semyonov firms produced the items of a traveling kit [*plate 71*] decorated with niello illustrations of architectural landmarks of old Moscow. One can imagine the warm wave of sentiment that swept over the Muscovite traveler who, having left his native city, fondly handled the articles in this kit. Though the items were factory-made, a substantial amount of handwork was invested in them. A silver cup and saucer with views of the Moscow Kremlin and St. Basil's Cathedral on Red Square [*plate 69*] may also have served someone as a wonderful memento of Moscow.

The firm of Carl Fabergé became the most famous jewelry firm of the late nineteenth and early twentieth centuries and saw its creations in demand far beyond the borders of Russia. This immense popularity was the result of the high artistic level of Fabergé products, the meticulous precision of their execution, and the irrepressible fantasy of creative ideas they embodied. Fabergé employed the finest craftsmen of the day, masters such as Mikhail Perkhin, Henrik Wigstrom and Julius Rappoport whose fame rivaled Fabergé's own.

One of the firm's regular clients was the Russian Imperial Court, whose commissions to a large degree fostered the reputation and glory of the firm. The most prominent of those commissions were for Easter eggs presented by tsars Alexander III and his son Nicholas II to Empresses Maria Feodorovna and Alexandra Feodorovna. The Armory collection includes ten Easter eggs, two of which appear in this exhibition. Each Fabergé egg had a "surprise" hidden inside, such as the wind-up model of the first Trans-Siberian express train inside a large silver egg made in 1900 [*plate 91*]. One of the most beautiful eggs contains an intricate model of the imperial yacht "Standart" [*plate 92*]. Its juxtaposition of transparent crystal, dark blue lapis lazuli, white, green, and dark blue enamel with diamonds, large pearls and gold creates a truly breathtaking effect. The intricate, minute reproduction of the yacht's details is awe-inspiring.

The Fabergé firm produced an amazingly wide range of items, including men's and women's jewelry, women's toilet accessories, silverware, writing implements, and cigar and cigarette cases. Fabergé masters devoted their skill and inventiveness to objects made not only of precious materials but also of more commonplace stones, often using "hardstone" materials from the Ural mountain regions. Carved hardstone figures of animals and fish, both whimsical and beautiful, enjoyed great popularity [*plate 86*]. In the second half of the nineteenth and the early twentieth centuries, when an interest in nature became an important direction in Russian jewelry art, brooches in the shape of apple tree branches [*plate 89*], lilies-of-the-valley [*plate 75*], dragonflies [*plate 87*], turtles [*plate 72*], assorted beetles [*plates 94, 95*], and butterflies became fashionable. Often they were executed in gold and silver and decorated with precious and semi-precious stones, pearls, or enamel.

The search for a national style in various branches of the arts, including jewelry-making, led nineteenth-century Russian craftspeople to turn to ancient, traditional forms and techniques. From those early models came decorative tableware crafted in the form of ladles, *bratina* winebowls, and small chests adorned with stylized ornament in older styles. In that same spirit grew a love for colored enamel over filigree, reminiscent of seventeenth century Muscovite patterns, and for motifs from folk legends and fairy-tales. This direction in art is vividly illustrated by such pieces as a Fabergé bratina winebowl and ladle [*plates 82, 93*], a teapot and milk-pitcher by the Saltykov factory [*plate 79*], and a sugar bowl and creamer made by the eleventh jeweler's *artel*, or guild [*plate 97*]. At the same time, as in Western Europe, the Russian arts saw a revival of European styles of previous centuries: Gothic, Rococo, and Classicism began to be reinterpreted according to national traditions. Examples of this are the address book and knife by the Fabergé firm [*plates 74, 106*].

By the end of the nineteenth century, the refinements of Modernism had begun to appear in jewelry design. The magnetic effect of cool tones and the white of silver and sparkling faceted crystal corresponded to the Modernist aesthetic and attracted forward-thinking customers. Large firms began to mass-produce a wide variety of crystal utensils set in silver. A pair of pitchers made by the renowned factory of Orest Kurliukov [*plate 103*] and the "Arctic" sauce boat [*plate 101*], recently acquired by the museum, are spectacular examples of this combination.

The second quarter of the twentieth century proved to be a barren period for the Russian jeweler's art, which was viewed as a moribund vestige of the past, obsolete in the new social order established by the 1917 Revolution. As a result, many accomplishments of the jewelers of the past were forgotten. This situation continued until the 1950s, when Russians, having recovered somewhat from the burdens of World War II, began to yearn again for beauty in everyday life, and professional artists willing to risk working in the medium of the jewelry arts re-emerged. Some worked at jewelry factories, while others operated independently in their own workshops. Today these two directions still exist, each with its own specific character and special points of interest.

Objects included in this exhibition that were manufactured in the leading jewelry factories of the former USSR are made of gold and precious stones. The Moscow Experimental Jewelry Factory is represented by an unusual brooch entitled the "Golden Knot Brooch" [*plate 109*], created in 1975 from a design by senior artist Alinia Petrova, and by an unusual memorial work in the form of a diamond crown by young designer Valerii Golovanov [*plate 114*]. Another modern master, Victoria Ignatiev, developed a beautiful set of jewelry while working in St. Petersburg at the Russian Gemstones Production Group [*plates 107, 108*].

The 1975 "Springtime" necklace is distinguished by its novel concept and splendid execution [*plate 110*]. This breathtaking piece is the work of a collective of craftspeople with different specializations working at the Sverdlovsk Jewelry Factory in the Ural Mountains. Designer Boris Gladkov, the factory's oldest artist, successfully conveyed in precious materials the sense of rebirth and new beginnings we feel each year as winter yields to spring and the natural world comes to life again.

The figurative principle is characteristic of the work of many Russian artists. This is exactly the departure point for works by the young Moscow artist Danil Chavushian. While still a student at the Moscow Higher Artistic-Industrial College, the industry's most prestigious institute of higher education, he visited the storerooms of the Armory in order to study closely the old works kept there. He later said that visit played a large role in his creative fate, strengthening in him the resolve to dedicate himself to the jeweler's art. Chavushian's favored technique is filigree, at which he has become a virtuoso.

Technical mastery, however, is not for him an end in itself but merely the means to an expressive goal. His Easter egg "The Golden Thread" [*plate 116*] hints at the secret of new life concealed inside the egg, a miracle hidden from view. Chavushian works in a small studio and executes his works entirely on his own.

Another talented contemporary artist, Mikhail Maslennikov, also works independently. In his elegant little box, inspired by Modernist style [*plate 115*], a number of materials have been successfully combined to delight the eye.

The pieces described above are among approximately 1,000 contemporary works of art of varying materials and specialties preserved at the Moscow Kremlin museum. Typical examples by a modern virtuoso are the intaglio "Chess Queens" [*plate 113*], the work of prominent St. Petersburg artist Pavel Potekhin. Potekhin is distinguished as a virtuoso stone-carver with the ability to produce an unexpected impression on the viewer.

Another St. Petersburg artist, Alexei Dolgov, specializes in carving on shell as he explores a wide variety of themes. The plaquette "Daughter of Peter the Great" [*plate 117*] was inspired by the Armory's collection of carriages and by events that transpired in the eighteenth century Kremlin. Like his forebears of earlier times, Dolgov collaborated with fellow craftsman Boris Sokolov, who provided the creative solution to framing Dolgov's pictorial conception.

From the ancient northern shores of the Black Sea to the opulent showrooms of cosmopolitan St. Petersburg, the Russian jeweler's arts have traveled a long path over the centuries, with many fruitful investigations and high artistic achievements along the way. Taking inspiration and example from the many cultures and peoples who left their stamp on the Russian character, the Russian jeweler has always at heart adhered to the native traditions of the Russian land. Today as in earlier epochs, the Russian jewelry artist looks to the past while moving forward into the future. The works presented in *Kremlin Gold* provide a vivid illustration of that path.

PLATES

Texts by: T.D. Avdusina [nos. 6–11]
I.A. Bobrovnitskaya [nos. 19–22, 44–47]
I. I. Vishnevskaya [nos. 23,28,27]
I.V. Gorbatova [nos. 116–122]
S.Y. Kovarskaya [nos. 71, 73–84, 88]
I.D. Kostina [nos. 55–70, 72]
M.V. Martynova [nos. 1–5, 12–14, 16–18, 24, 25]
E.A. Morshakova [nos. 15, 41]
T.N. Muntyan [nos. 85–87, 89–93, 95–101, 105–111, 113–115, 123]
V.M. Nikitina [no. 94]
L.N. Peshekhonova [nos. 124–140]
E.A. Yablonskaya [no. 26]

Scientific editor: N.S. Vladimirskaya
Russian text editors: E.B. Gusarova, I.A. Sterligova

This flexible chain bracelet is made of braided gold wire. Two serpent heads, cast in gold and decorated with colored glass, form the clasp. Although in evidence since ancient times, braided bracelets such as this were far less common than their solid metal counterparts. Found in 1927 with the Suzdhal treasure trove, this bracelet demonstrates ancient artisans' supreme skill in twisting, plaiting, and drawing out pliable gold wire – a skill that would continue to mark the finest creations of Moscow's craftsmen in later centuries.

1 *BRAIDED GOLD BRACELET*

Northern Black Sea region
4th–5th century
Gold, colored glass
Gold casting and braiding
Length: 22 cm

This decorative gold collar, or *grivna*, weighing more than two pounds, consists of a circular medallion inlaid with cherry-colored glass and mounted on a hollow tube of gold. The style of this grivna decoration, which originated in ancient Rome, is now known as North Black Sea cloisonné mosaic. Part of the Suzdhal treasure found in 1927 near Kursk in Russia's northern Black Sea region, this collar stems from a time of migration of peoples through the future lands of Russia, foremost among them the Huns, who conquered this area in the fourth and fifth centuries on their march into western Europe.

2 *GOLD GRIVNA*

Northern Black Sea region
4th–5th century
Gold, colored glass
Gold casting, mosaic inlay
Diameter: 22 cm

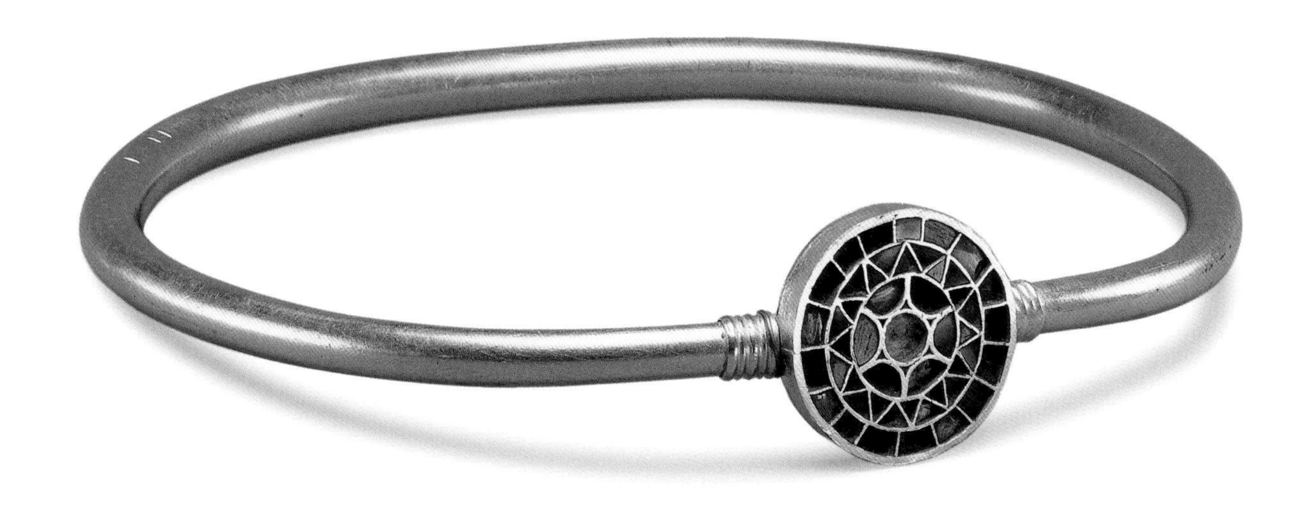

This octagonal reliquary was designed as a domed shrine, with hinged doors that open to reveal a sarcophagus used to contain a reliquary object. A Greek poem inscribed on one panel identifies the reliquary as a replica of the shrine marking the grave of Saint Dmitry in the Basilica of Dmitry at Thessalonika, Greece. Guarding the doors are the soldier-saints Nestor and Lup, Dmitry's beloved companions, while on the other side of the shrine appear the figures of the Byzantine Emperor Constantine Duk and his wife Yevdokia. The graceful proportions of these figures reflect the classical heritage that filtered into Russia through contact with Byzantium.

The aristocratic language of the inscribed poem suggests that the shrine may have been created as a gift for the Emperor, who is shown with the Empress in a marriage ceremony conducted by Jesus. Taken to Russia in the twelfth century and installed in the Cathedral of Dmitry in Vladimir, the shrine eventually became part of the imperial collection of reliquaries. In the eighteenth century it was placed in the Moscow Kremlin's Annunciation Cathedral.

3 *RELIQUARY SHRINE*

Byzantium
11th century
Gold, silver
Gold chasing and engraving, silver gilding
Height: 15 cm

This icon, a cameo relief carving of the image of Christ, was executed in banded, multicolored jasper and mounted in a rounded gold frame trimmed with filigreed denticles. At top and bottom the frame holds rectangular agates, while the lower corners are adorned by two faceted stones. The frame is also set with eight bezel-mounted stones, four of which are cabochons of jasper and carnelian.

Common in ancient times, the technique of hardstone carving was redeveloped in the Byzantine world during the tenth century. Because the carving of these hardstones often required more than one year of a craftsman's labor, they were only produced on commission from wealthy, high-ranking officials. The Kremlin Armory Museum owns one of the world's most important collections of Byzantine carvings in lapis, chalcedony, and agate.

4 JASPER CAMEO ICON: "CHRIST GIVING BLESSINGS"

Byzantium
11th century
Gold, jasper, carnelian agate, emerald, and almandine garnet
Gold chasing and engraving, gem carving and polishing
Height: 12 cm

1000s

This silver signet ring, one of fifteen in a treasure trove found on the Kremlin grounds, features an engraved eagle standing against a dark background. The eagle's right wing is raised, the left wing appears to be lowered to the ground, and the head is turned to one side as if the bird is looking back over its left shoulder. This central image is framed by four engraved decorative elements that are repeated on the sides of the ring.

Slightly smaller than its counterpart in plate 5, this silver signet ring also features an engraved eagle. Though the design seen here is far more simple, the elements of the central image and the eagle's pose are the same. In contrast to the eight curved surfaces that surround the central image of the larger ring, this piece features a simple line engraving incised upon a single circular planchet.

6 *SILVER SIGNET RING*

Old Russia
11th–13th century
Silver
Silver engraving and gilding
Diameter: 2.4 cm

5 *SILVER SIGNET RING*

Old Russia
11th–13th century
Silver
Silver engraving
Diameter: 2.8 cm

Reconstructed from depictions in historical church manuscripts, these sixteen vertical ornaments, fastened parallel to one another, form a sort of headband to embellish a woman's head covering. Medallions known as *kolts* hang from either side. Both the medallions and these ornaments were part of the treasure trove of Tula.

This spur-shaped medallion, or *kolt*, found in the treasure trove of Tula, was constructed using a variety of silversmithing techniques. Meticulously detailed filigree and granulation work can be seen in the more than five thousand "seeds" of silver that decorate the surfaces. Each seed is surrounded by a filigree band. This medallion hung from one side of a woman's headdress, balanced by a similar piece on the other side, as seen on the facing page.

8 *SILVER KOLT*

Old Russia
12th century
Silver
Silver casting, filigree, and granulation
Length: 11 cm

1100s

7 *SILVER HEADDRESS DECORATION*

Old Russia
12th century
Silver
Silver casting and filigree
Length of each: 6 cm

This double-hinged, cylindrical, cuff-styled bracelet, a typical piece of jewelry from pre-Mongolian Russia, would have been worn on the forearm over wide, heavy sleeves. The engraved decoration – fanciful birds and griffins on the upper portion, vines and leaves encircling the lower section – recalls similar motifs found in the decoration of the cathedral at Vladimir, where this piece is thought to have been made. Each decorative section is framed with engraved and granulated silver work. The simple and restrained, yet powerful and elegant forms of this piece display the small-scale monumentality that marked many of the crafts of Old Russia.

9 *SILVER BRACELET*

Old Russia
12th–13th century
Silver, niello alloy
Silver casting and engraving, niello
Diameter: 7.5 cm

1100s

This icon is an early example of cloisonné-style enamel work, which was first developed in Constantinople in the sixth century. Byzantine cloisonné enamel was famed for the variety of colors used and the high quality of the workmanship. Here the image of Christ, who destroys the gates of Hell while holding Adam and Eve by the hand and leading the righteous to Paradise, is depicted in various colors of enamel inlaid in an engraved sheet of gold, with gold wire work separating the colors.

The Byzantine cloisonné enameling technique, widely used in church icons, was also employed to decorate jewelry, clothing, armor, plates, and sarcophagi.

10 RELIQUARY ICON: "DESCENT INTO HELL"

Byzantium
12th century
Gold, silver, enamel
Gold engraving, silver casting, cloisonné enameling
Height: 9.5 cm

This collar-like shoulder mantle, or *barmy*, was a sign of royalty worn by a grand prince or other high-ranking official during Russia's pre-Mongol period. This barmy features five large, circular medallions, each suspended between oblong beads of openwork filigree. The cloisonné enameled portraits of the saints are accompanied by engraved inscriptions identifying them as Mary, Mother of God (center), St. Irina (left), and St. Barbara (right). The medallions are accented throughout with more than fifty bezel-mounted amethysts, garnets, spinels, and other gemstones. Like other Old Russian works in the Armory collections, this barmy displays the simple, massive forms typical of Russian jewelry and decorative art made prior to the Mongol invasions.

1100s

11 *GOLD BARMY*

Ryazan, Old Russia
12th–13th century
Gold, amethyst, garnet, spinel, pearl, glass, enamel
Gold filigree, engraving, and granulation, cloisonné enameling
Collar length: 40 cm

Discovered in 1822 on the site of Ryazan, the major jewelry-producing center of pre-Mongolian Russia, this medallion, or *kolt,* was part of a decorative headdress. Weighing close to one pound even with a hollow core for holding fragrances, this kolt is exceptional among objects of its kind. The rim of the kolt is covered with filigree and bezel-mounted precious stones.

The figural decoration on the face of the kolt evokes the nearly continuous fratricidal wars of succession which plagued Old Russia. Symbolized together in one composite image are the brothers Boris and Gleb, who were murdered in the eleventh century by their older brother Svyatopolk in his quest to eliminate contenders for his throne. The tragic early death of the young dukes contributed to their later canonization: both were declared saints in 1171.

On the reverse of the Ryazan kolt is a nearly flawless piece of rock-crystal quartz mounted in gold and surrounded by pearls. The stones, set symmetrically around the perimeter, include amethyst, almandine, sapphire, and emerald.

12 GOLD KOLT

Ryazan, Old Russia
12th–13th century
Front: Gold, pearl, jasper, garnet, enamel
Back: Gold, pearl, quartz, garnet (almandine), sapphire, emerald
Front: Gold chasing and filigree, cloisonné enameling
Back: Gold chasing and filigree
Diameter: 12.5 cm

This delicate openwork bracelet was part of the treasury of Old Ryazan, which was destroyed by Mongol invaders in 1237. Presumably buried for safe-keeping in advance of the attack, this treasure was unearthed in 1822 by a farmer plowing his fields. Unlike traditional, more massive Old Russian bracelets, this piece was made with flattened, twisted, and rounded wires of gold of varying thickness and would have been worn loose on the wrist, rather than over a heavy sleeve. This piece of fine jewelry may have belonged to the wife or daughter of a grand prince or high-ranking official.

13 *GOLD FILIGREE BRACELET*

Ryazan, Old Russia
12th–13th century
Gold
Gold filigree and chasing
Diameter: 6.3 cm

1100s

Made by master artisans in the principality of Vladimir-Suzdhal, this reliquary cross exemplifies a variety of goldsmithing techniques and a form of decoration unique among thirteenth-century gold jewelry. The gold surface of the cross, its rounded ends familiar from Russian icon paintings of this time, is covered almost entirely by finely-wrought heart-shaped designs. Made from thick, twisted gold wires accented with granulated beads of gold, this impressive ornamentation has no close analogy, although the elegance of the filigree echoes ancient Roman examples.

Unlike western European reliquaries, which often expose the sacred relic, Russian reliquaries completely conceal their precious contents. This example, connected by a hinge to a rectangular piece that hung from a chain, belonged to the Kremlin's Annunciation Cathedral before being transferred to the Armory Museum in 1922.

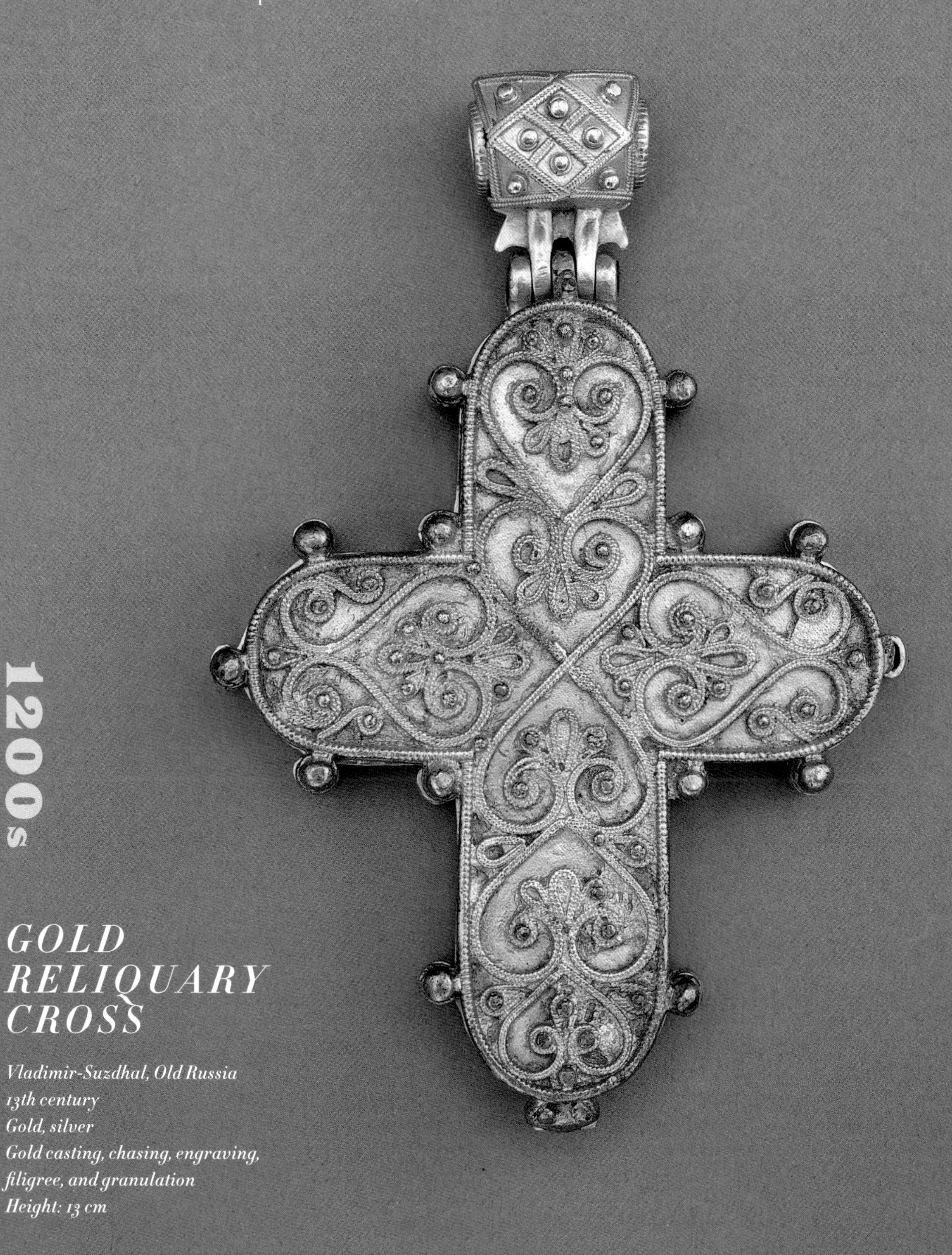

1200s

GOLD RELIQUARY CROSS

Vladimir-Suzdhal, Old Russia
13th century
Gold, silver
Gold casting, chasing, engraving, filigree, and granulation
Height: 13 cm

This gold crown was part of an *oklad*, or icon cover, that was made for the "Mother of God of Bogolubskaya" icon in the late thirteenth or early fourteenth century. The trefoil domes of each panel in the crown and the technique used to mount the stones recall styles originating in the twelfth century, while the filigree and granulation echo such pre-Mongolian pieces as those in the Ryazan treasure hoard. [*plates 11, 12, 13*]

The oklad that this crown once decorated is known from detailed descriptions in fifteenth century documents. The crown, however, is all that survives – the rest of the icon cover vanished during the advance of Napoleon's armies against Moscow in 1812.

15 *ICON CROWN: "MOTHER OF GOD OF BOGOLUBSKAYA"*

Moscow
13th–14th century
Gold, sapphire, garnet, turquoise, emerald, pearl
Gold chasing and filigree, pearl drilling
Length: 30 cm

1200s

Novgorod, Moscow's great cultural and political rival until the late fifteenth century, was Old Russia's primary center for carved sculpture in miniature. The precisely executed details in this reliquary attest the skill of Novgorod's craftsmen in carving materials such as slate, shale, steatite or, as seen here, wood. The image on this folding icon, a "Crucifixion of Christ with Attending figures," was probably inspired by a painted original. On the inside panels of the doors are carved the images of prophets, scenes from the life of Christ, and selections from the psalms of King David. The size of this icon and the design of the clasp suggest that it was designed to travel as a personal effect that could also be worn suspended from a chain.

16 CARVED WOOD AND GOLD RELIQUARY ICON

Novgorod
15th century
Gold, silver, wood
Gold gilding and filigree, silver engraving and chasing, wood carving
Length: 8.8 cm

1400s

This *zion*, or reliquary, one of four commissioned by Ivan III in 1486 for Moscow's Annunciation Cathedral, is based on an ancient Russian form first mentioned in eleventh-century manuscripts. Although the original purpose of the zion is unknown, it is thought to have served as a liturgical vessel in the eleventh and twelfth centuries. During the fifteenth century zions were used to contain gifts given on ceremonial occasions, and by the seventeenth century had found their way into special religious ceremonies in Moscow's Assumption Cathedral. With their ultimate source in Jerusalem's own Annunciation Cathedral, and known as "Jerusalems" before the seventeenth century, zions typically reflected the church architecture of the Old Russian cathedral cities where they were made. The structure of the zion echoes that of an actual Russian Orthodox cathedral, in which tiers of domed cupolas, called *kokoshniks,* symbolize the heavenly city of Jerusalem. The zion seen here, with three levels of kokoshniks ascending to a central "onion" dome, reflects the authentic architecture of Moscow's churches, while the proportions of both the building and the figures suggest the influence of the European gothic style on the native Russian designer.

The original of this small zion disappeared in 1918 during an robbery of the Russian Patriarch's Palace. This replica, produced in 1913 by the Imperial Historical Museum, was later transferred to the Armory Museum.

1486

17 *ASSUMPTION CATHEDRAL ZION*

Moscow, 1486
replica, 1913
Silver, copper, gold, niello alloy
Gold gilding and chasing, niello
Height: 63 cm

This gold icon cover, or *oklad*, was originally made as a frame to fit over a painted icon depicting the Madonna and Child. Rubies, emeralds, sapphires, and pearls serve as centerpieces for the floral designs in thin filigree filled with multi-colored enamel that cover almost the entire surface. Drops of gold in the dark red enamel enhance this oklad which, though traditional in design, is one of the finest known examples of sixteenth-century Russian goldsmithing.

Although, like most Russian arts of this period, this oklad carries no stamp to indicate the time or place of its manufacture, clues in the decoration of the frame suggest its date of origin. Ten niello images in the medallions around the frame depict the namesake saints of Ivan the Terrible's family. The image of the soldier-saint Fedor Stratilat symbolizes Ivan's third son Fedor, born in 1557, while that of St. Anastasia represents the protector of Anastasia Romanov, Ivan's first wife. Because Ivan had married again by 1561, it is assumed that this oklad was commissioned before that marriage and after the birth of Fedor.

"Odighitria" is a Greek term meaning "one who shows the way."

18 ICON COVER: "OUR LADY OF ODIGHITRIA"

Moscow Kremlin Workshops
1557–60
Gold, ruby, sapphire, emerald, tourmaline, pearl
Gold chasing, filigree, and engraving, enameling, niello
Length: 47.5 cm

1557

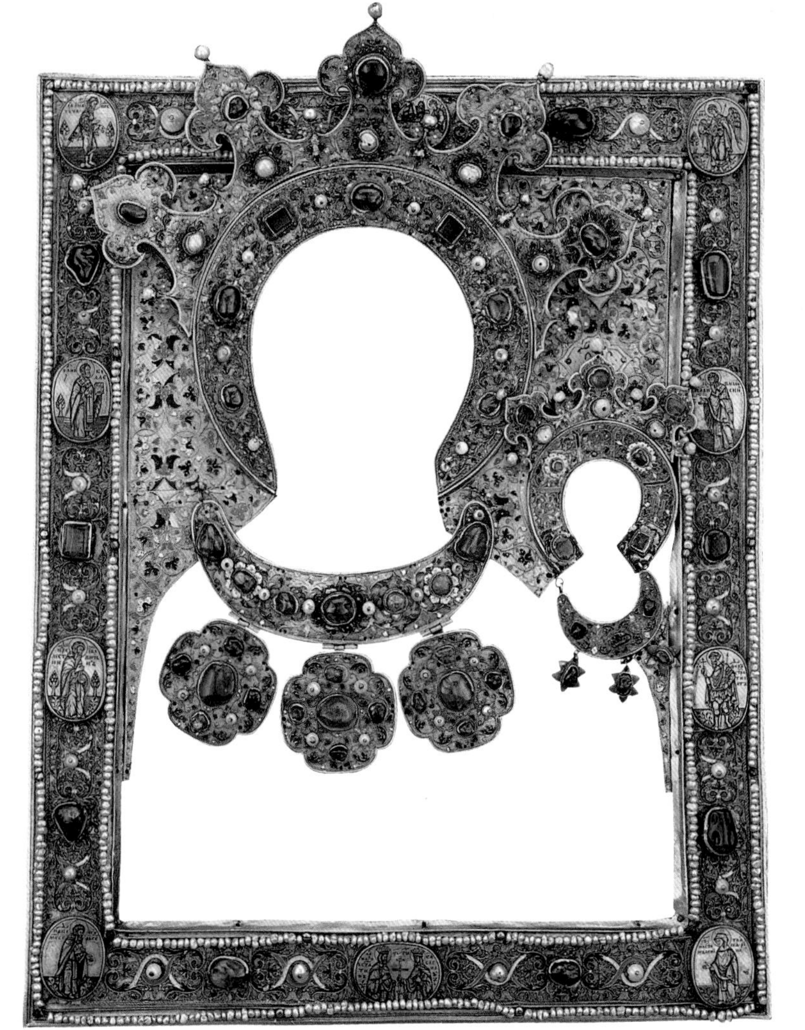

This sumptuous reliquary icon was presented to the Archangel Cathedral of the Moscow Kremlin by Tsarina Irina Godunova, the daughter-in-law of Ivan the Terrible and sister of Boris Godunov. The inscription on the face states that the reliquary was commissioned in honor of the tsarina by her husband, Tsar Fedor Ivanovich, Ivan the Terrible's son and successor and the last tsar of the ancient Rurik dynasty. The reverse carries a niello engraving of St. Irina. The chased gold image of the Madonna and Child is surrounded by an opulent frame studded with rubies, emeralds, sapphires, and pearls

In striking contrast to his tempestuous father, who claimed his son had been born for the monastery and not the throne, Fedor was a weak, sickly, yet kind-hearted ruler. His profoundly religious nature may account for his choice of this icon as a present for his wife. Historical sources claim that Fedor commissioned the reliquary to commemorate the appointment in 1589 of Jeremy of Constantinople as the first patriarch of the Russian Orthodox Church. The relics reportedly brought by Jeremy to his new tsar in Moscow may have been housed within.

19 *RELIQUARY ICON OF TSARINA GODUNOVA*

Moscow Kremlin Workshops
1589
Gold, ruby, emerald, sapphire, pearl
Gold chasing, engraving, niello
Height: 11.8 cm

1589

MP ΘY
IC XC

This *panagia*, a jeweled pendant containing an icon worn by high-ranking Russian clergymen, features a cameo carving of the "angel of the desert," John the Baptist, patron saint of Ivan the Terrible. With one hand grasping a staff that blooms into a cross, the saint carries his head in the other hand to symbolize his terrible death. Surrounded by pearls, precious stones, and enamels, the cameo is carved from a single piece of three-colored agate. Although the unique structure and color of the agate suggest a source in Italy, with which Russia had frequent contact in the sixteenth century, the eastern style of the carving is characteristic of Byzantine, Russian, and Balkan art.

Probably created for Ivan the Terrible, this panagia twice disappeared from the tsar's treasury and twice reappeared in Moscow. First lost during the Swiss-Polish invasions of Moscow in the early seventeenth-century "Time of Troubles," it was returned to Moscow to rest safely until being sent to St. Petersburg in the eighteenth century. Disappearing again from St. Petersburg's Winter Palace, the panagia turned up in a Moscow jeweler's shop in 1927, and was finally returned to the Moscow Kremlin.

20 "ANGEL OF THE DESERT" PANAGIA

Moscow Kremlin Workshops
16th century
Gold, silver, diamond, ruby, tourmaline, pearl, agate (cameo), niello alloy, enamel
Gold chasing and filigree, niello, enameling, cameo carving
Height: 16 cm

This boatshaped dipper, or *kovsch*, once belonging to Tsar Boris Godunov, was forged from a single piece of gold. It is similar in form to the wooden dippers that were used as drinking vessels in Russia as early as 2000 B.C., which often took their shape from waterfowl such as wild ducks, geese, and swans. By the end of the fifteenth century, the kovsch had been transformed from a utilitarian vessel into a presentation piece given as an award or used to commemorate special occasions. Red mead, an alcoholic drink similar to wine, was served in a gold kovsch, while white mead was served in a silver one. The interior of this example bears an engraving of the imperial insignia, a double-headed eagle wearing a crown. The tsar's imperial titles are inscribed in niello on the sides of the kovsch. Although an inventory of Boris Godunov's property lists numerous examples of the kovsch, only the one shown here is known to have survived.

In Russian tradition relatives expressed gratitude or affection by passing precious objects from hand to hand. This kovsch was listed in the property of the *boyar* (noble) Nikita Romanov, who probably received it from his cousin and Godunov's eventual successor, Tsar Mikhail Romanov. With no offspring of his own to inherit it, Nikita's property, including this kovsch, reverted to the imperial treasury.

21 *GOLD KOVSCH*

Moscow Kremlin Workshops
16th century
Gold, niello alloy
Gold forging, chasing, and engraving, niello
Length: 18.5 cm

Created by Russian master goldsmiths Danila Osipov and Tretyak Pestrikov, this gold censer, or *kadila*, was presented by Tsar Mikhail Romanov to the Trinity Monastery of St. Sergius in 1616, a gesture commemorated by the niello inscription at the bottom of the vessel. Although different types of kadila were known in Old Russia, from the fifteenth through the seventeenth centuries their form replicated the architecture typical of Russian Orthodox churches, which feature domed cupolas topped by an orthodox cross.

Tsar Mikhail's gift reflects the close and ancient ties that bound the tsars to the powerful Russian Orthodox church. Established in 1345 by Sergei Radonezhsky and his brother Stefan, the Trinity Monastery was the spiritual center of the Russian state and played a major historical role as the tsar's bulwark in times of religious or political foment. Over several centuries a priceless collection of art was amassed at the monastery, donated by the highest-ranking families of Russia's grand dukes and tsars. This kadila, considered the pearl of the collection, was transferred to the Armory Museum in 1928.

22

GOLD KADILA

Moscow Kremlin Workshops
1616
Gold, sapphire, emerald, tourmaline
Gold chasing and engraving, niello
Height: 31.5 cm

This presentation *kovsch*, or dipper, which weighs almost two kilograms, was forged from a single piece of gold – evidence of Russia's immeasurable mineral resources. Presented to Tsar Mikhail Feodorovich Romanov, the founder of the Romanov dynasty, by his mother Maria Ivanovna, it is accented with sapphires and rubies, trimmed with freshwater pearls, and enhanced by fine niello infill. Originating in the sixteenth century, this form of lavishly decorated kovsch, with a wide, flat, shallow bowl, was used during the seventeenth century to recall the grandeur of that earlier age. Kept with the tsar's main treasury and brought out for special occasions only, objects like this kovsch were used to impress visiting dignitaries with the immense wealth and splendor of the Russian tsar. Historic documents relate that, at a dinner in the Kremlin's Cross Room in 1571, the patriarch of the Russian Church was served red honey in three kovschs "highly decorated with pearls and stones," such as the one shown here.

23 *GOLD KOVSCH OF TSAR MIKHAIL ROMANOV*

Moscow Kremlin Workshops
1618
Gold, silver, ruby, sapphire, emerald, pearl
Gold forging, chasing, and engraving; niello
Length: 30 cm

Ceremonial plates and tableware, derived from the simple utilitarian forms of medieval Russia, are among the most numerous and varied of the objects in the Armory collections. This *bratina* illustrates one of the most common types of ceremonial drinking vessel, the loving cup, which was passed hand-to-hand as each guest drank. Introduced in the later sixteenth century, bratinas ranged from massive vessels the size of huge mixing bowls to small elegant pieces such as this one, made expressly for a woman. This bratina is decorated with enamel designs and a niello inscription.

More significant than the size or shape of the bratina was its inscription, which may have carried the owner's name, a dedication, a homily, or an expression of love and affection. The bratina was often given by husbands to their wives to mark special occasions, as in this example, whose inscription tells us that it belonged to Yevdokia, the wife of Tsar Mikhail Feodorovich Romanov.

24 GOLD BRATINA WITH COVER

Moscow Kremlin Workshops
1626–45
Gold, enamel
Gold chasing, engraving, and enameling; niello
Height: 6.6 cm

1626

This *kolchan*, or quiver, is part of the ceremonial full dress armor made for Tsar Mikhail Feodorovich Romanov and worn exclusively for reviews of the imperial troops during military parades, for ceremonial receptions, and for audiences with foreign ambassadors. The tsar's complete outfit included helmet and body armor, shield, mace, sword, bear spear, and the *saaddak*, or bow and arrow set, which was carried in the kolchan. The kolchan seen here is lavishly decorated with gold, enamel, diamonds, sapphires, rubies, and emeralds, and includes the imperial double-headed eagle insignia at the top.

25 *GOLD KOLCHAN OF TSAR MIKHAIL ROMANOV*

Moscow Kremlin Workshops
1627–28
Gold, silver, diamond, ruby, sapphire
Engraving, chasing, enameling
Length: 47.5 cm

The bowl of this wine tasting cup, or *charka*, is carved from a single crystal of quartz from Russia's Ural mountains. The gold rim and pedestal are accented with rubies and emeralds. Inscribed on the rim and accented in niello is the name of Tsar Mikhail Romanov, while on the bottom of the handle a second inscription explains that Mikhail's wife, the Tsarina Yevdokia, presented this cup to her young son Ivan, who died at the age of six. Charkas were used to drink wine, mead, or other alcoholic drinks on special occasions.

Vessels like this charka demonstrate the seventeenth-century Russian jeweler's delight in combining rare materials of starkly varied character to achieve bravura effects of bold contrast. This one is referenced in an inventory of Tsar Mikhail's treasury among lists of ceremonial gold, silver, and crystal vessels that incorporate semi-precious materials such as agate, carnelian, quartz, jasper, and serpentine.

26 *GOLD AND QUARTZ CHARKA*

Moscow Kremlin Workshops
1636
Gold, quartz (rock crystal), emerald, ruby, niello alloy, enamel
Gold engraving, niello, enameling
Diameter: 6.5 cm

1636

This Assumption *panagiar*, or altar plate, was used to hold the communion wafers offered to Mary, the Mother of God, during special presentation rituals held in the church or monastery and at the tsar's official state dinners. In this ornate example, the polished gold plate is held aloft by four golden angels with enameled halos and silver wings. The angels, in turn, stand on the backs of stylized lions, each of which holds the tail of another in its mouth.

This panagiar recalls the style of an earlier example made in 1435 for Novgorod's St. Sophia Cathedral. However, it is more highly detailed and decorative, as is characteristic for mid-seventeenth century applied arts from Moscow. Hellenic elements in the panagiar also suggest that it might have been placed in Moscow's Assumption Cathedral by Patriarch Nikon, who attempted to introduce Greek Orthodox practices into the Russian church in the mid-seventeenth century. The custom of offering the communion wafer to Mary was also borrowed from the Greek church.

27 GOLD PANAGIAR

Moscow
17th century
Gold, silver, enamel
Gold gilding; silver chasing, engraving, and filigree; enameling
Height: 31.5 cm

This life-sized, chased, sheet-gold effigy was made as a sarcophagus cover for the coffin of Tsarevich Dmitry, youngest son of Ivan the Terrible. Dmitry died as a child in 1591, murdered, legend claims, at the order of Boris Godunov. He was canonized in 1603. The effigy is one of only three such objects to have survived in Russia, all thought to be from the workshop of master jeweler Gavriil Ovdokimov.

The tsarevich is depicted with the gentle features of a young boy, his head surrounded by a halo of gold and his robes accented with numerous gemstones. The scrollwork on the surrounding frame is interspersed with the images of the Romanov's patron saints. The rhythmic chasing of the patterns covering the frame hints at eastern origins, while the naturalistic approach to the face and figure suggests influences from the Italian Renaissance.

This cover and a silver shrine were commissioned by Tsar Mikhail Romanov thirty years after Dmitry's death. The tsar ordered the saint's remains transferred to the Kremlin's Annunciation Cathedral, where they lay safely until Napoleon's invasion of the city in 1812. Although the shrine itself has disappeared, the golden cover was carefully hidden away and survived.

28 *GOLD SARCOPHAGUS COVER*

Moscow Kremlin Workshops
1630
Gold, silver, amethyst, emerald, ruby, sapphire
Gold gilding, silver chasing and casting
Length: 157 cm

Miters of the type pictured here, with a rounded crown divided into four decorative segments, began to appear in the early seventeenth century. This ornately decorated piece is encrusted with precious gems and pearls that frame gold medallions bearing the enameled images of holy figures. Those just above the miter's golden rim depict Fedor, David, and Constantine, who were known as miracle-workers in the city of Yaroslavl. Above them are the engraved images of the Muscovite bishops Peter, Alexei, John, and St. Nikolai, while in a medallion at the very top of the crown is an enameled Russian Orthodox Trinity.

A niello inscription engraved on the gold rim provides an unusually precise date for this miter: "On August 15th in the summer of 1634, by the command of the Great Ruler, Tsar, Grand Duke and Autocrat of all Russia Mikhail Feodorovich Romanov, this miter was made in the Mother of God Church of Rostov in the memory of the great Filaret, Patriarch of Moscow and all Russia." In 1834 an extensive restoration of the miter was performed to replace missing stones and the velvet support.

29

BISHOP'S MITER

Moscow Kremlin Workshops
1634, restoration 1834
Gold, silver, diamond crystals, diamond, emerald, ruby, pearl, enamel, niello alloy, velvet
Enameling, niello
Height: 22 cm

1634

This engraved silver container was used to chill flasks of wine for special occasions. Engraved floral patterns adorn the sides and cover. Chased belts around the container carry the image of oaken casks. This vessel displays extensive evidence of western influence: the handles were probably brought from Western Europe and applied to the vessel, while the pine cone handle is a motif common to Dutch and German silverware. The engraved garland designs with tulips and birds, also a western detail, were probably added in the late seventeenth century, in a display of the "flower style" popular from the 1650s to the 1670s. The inscription on the rim states that this dish was once owned by the *boyar* (noble) Vasily Streshnev, who for many years was the head of the Moscow Kremlin workshops.

30 *SILVER CHILLING DISH*

Moscow Kremlin Workshops
1646
Silver
Silver casting, chasing, and engraving
Height: 31.7 cm

1646

Once owned by Patriarch Josef II, who led the Russian Orthodox Church from 1667 through 1672, this pectoral insignia, or *panagia,* features a twelfth-century chrysoprase carving depicting the "Assumption of the Mother of God." The gold setting was designed and executed in 1671 by the Moscow jeweler Mikhail Yakovlev. Accent stones include rubies, emeralds, sapphires, and tourmalines. The central pendant stone is a Ceylon sapphire.

The sculpture decorating this panagia, created in the twelfth century, is unique among Byzantine cameos in presenting not the typical single figure or portrait but, rather, a scene including fifteen individuals – evidence that a master carver produced this piece. The soft green color of the chrysoprase cameo is enhanced by the rich blend of colors in the gem-studded frame that surrounds it.

31 *PANAGIA: "ASSUMPTION OF THE MOTHER OF GOD"*

Moscow Kremlin Workshops
1671
Gold, chrysoprase, emerald, ruby, sapphire, tourmaline, pearl, enamel
Gold filigree, enameling
Diameter: 9 cm

1671

This icon depicts the traditional Russian Orthodox Trinity, a composition of three seated figures inspired by the Biblical story of the three angels who appeared to Abraham under an oak tree. Similar images appearing in fifth- and sixth-century Byzantine mosaics account for the later adoption of this convention by Russian icon painters. In this piece, Abraham (holding a plate of breads), his wife Sarah, and Abraham's sacrificial calf accompany the three holy figures. Naturalistic details in the background – the splendid palace seen behind Sarah, the patch of deep blue sky, the graceful trees breaking the horizon – relieve the static presentation and suggest spatial depth.

The gold frame and cover are mounted over the icon, which is painted in egg tempera on wood and is attributed to the School of Simon Ushakov.

Supposedly commissioned by Tsar Fedor Alexeivich Romanov, this icon was later listed among the personal possessions of Tsar Nicholas II and was transferred from St. Petersburg to the Moscow Kremlin after the October Revolution of 1917.

32 GOLD ICON COVER

Moscow Kremlin Workshops
late 17th century
Gold, diamond, wood, fabric, egg tempera paint
Gold chasing and engraving, woodcarving, enameling
Length: 28.2 cm

This stole, the surviving remnant of a larger set of clerical vestments, was worn over the shoulders by Russian Orthodox bishops during services. The superb workmanship and the wealth of precious materials employed suggest that this stole was commissioned by the imperial court, for only the tsar's treasury could provide such a quantity of gold, gems, and pearls. The thousands of gold sequins sewn to the black velvet stole create the impression of a forged gold surface covered by floral decorations of tiny pearls outlined by strands of larger pearls. Diamonds, emeralds, and rubies make up the blossoms on each stem and the cross that accents the center of the stole. Ornate objects such as this stole were often destroyed in later attempts to reclaim the priceless materials they contained.

1600s

33 *PEARL STOLE*

Moscow Kremlin Workshops
late 17th century
Gold, silver,
diamond, pearl
Gold casting and chasing,
embroidery, weaving
Width: 76 cm

This gold and silver altar cross is designed in the traditional Russian Orthodox form: a single vertical shaft crossed by two horizontal bars and one diagonal bar to represent the Christian concept of a triune God. This cross, outlined in rubies and emeralds and decorated with enamel, carries a Cyrillic inscription in niello indicating that it was commissioned for the Kremlin palace churches by Tsar Fedor Alexeivich, the oldest brother of Peter the Great. Similar crosses were used on the altars of virtually every church in Russia and in the fifteenth century often served as reliquaries.

The sickly Fedor, who assumed the throne at the age of fourteen and held it for six short years before his death, was a deeply religious youth who made frequent, lavish donations to churches, monasteries, and cathedrals. His characteristic gifts were precious liturgical objects such as gospel covers encrusted with gems and splendid crosses like the one pictured here.

1677

34 GOLD ORTHODOX ALTAR CROSS

Moscow Kremlin Workshops
1677
Gold, silver, ruby, emerald, diamond
Gold gilding, silver chasing and engraving, niello enameling
Length: 35.5 cm

Pictured here is one of two crowns made for the brothers Ivan and Peter who, for a time, ruled Russia together as Tsar Peter and Tsar Ivan Alexeivich. After Ivan's death in 1696, Peter the Great continued to reign alone as tsar until his own death in 1725. This crown, made for the older brother Ivan, contains hundreds of large diamonds mounted in gold and silver and arranged in a pattern that alternates floral designs with the imperial double-headed eagle insignia. At the top of the crown a diamond cross is mounted on a transparent piece of uncut tourmaline. With diamonds and gold competing for dazzling effect and diamond-encrusted medallions set at varying angles to intensify the reflection of light, this crown exemplifies the regalia used to impress foreign dignitaries, as well as the Russian people, with the power and splendor of the tsar. The style of this sable trimmed crown originated in the thirteenth century and continued to influence the design of tsarist crowns for hundreds of years.

35

DIAMOND CROWN OF TSAR IVAN ALEXEIVICH

Moscow Kremlin Workshops
1682
Gold, diamond, tourmaline, sable, enamel
Gold chasing and casting, enameling
Height: 29.5 cm

Ceremonial plates in precious metals, often enormous in scale and ornately decorated, were produced locally for the court or received as gifts from foreign ambassadors. This presentation piece from the Kremlin Workshops is a striking example of state-of-the-art goldsmithing at the end of the seventeenth century. The enamel work is accented by rubies and emeralds.

1694

36 *GOLD PLATE*

Moscow Kremlin Workshops
1694
Gold, ruby, emerald
Gold casting and chasing, enameling
Diameter: 15.2 cm

This *bratina,* or loving cup, is decorated with clusters of diamonds, rubies, and emeralds mounted in ten oval medallions that protrude from the side of the bowl. Ten corresponding medallions decorate the lid, while a row of alternating rubies and emeralds outlines the outer rim of the cover. The Cyrillic inscription indicates that Peter the Great gave this cup to his son, Tsarevich Alexei Petrovich, on his fourth birthday.

Alexei, the child of Peter's first marriage to Yevdokia Lopukhina, was eventually scorned by Peter as weak-willed and unfit to rule. Alienated from his father and embittered when Peter banished Yevdokia to life in a monastery, the young Alexei finally sided with the powerful *boyars* who opposed Peter's reforms. He fled Russia in 1716 but was brought back at his father's order and executed for treason.

37 *GOLD BRATINA WITH LID*

Moscow Kremlin Workshops
1694
Gold, diamond, ruby, emerald, enamel
Gold casting and chasing, enameling
Height: 10.4 cm

Ryasny, pendant ornaments worn over the ear and used in Old Russia to embellish a married woman's headdress, were also often attached to icons of Mary, the Mother of God. Two of four pairs owned by the second wife of Tsar Mikhail Romanov were donated by her to adorn the "Mary of Vladimirskaya" icon. Made of pearls, turquoise, and almandine fastened to a thin sheet of gold for added strength, the ryasny seen here were commissioned by Tsar Mikhail and his mother and given to the Ipatievsky monastery in Kostroma to decorate the "Mary of Feodorovskaya" icon.

1600s

38 PEARL RYASNY

Moscow Kremlin Workshops
17th century
Gold, silver, turquoise, pearl, niello alloy
Gold chasing, bead work, niello
Length: 38 cm

The central image on these earrings, a pair of birds, is a traditional Russian symbol of renewal. The enameled birds are perched on a bezel-mounted, faceted emerald beneath an arch made of rubies and gold. Below the birds, three pendant pearls hang from a golden swag decorated with white enamel flowers.

Pendant earrings were traditional ornaments for sixteenth- and seventeenth-century Russian women. Those shown here are designed as stylized birds decorated with white and green enamel and accented by rubies and emeralds. Each piece contains a drilled, ruby-colored, oblong bead surrounded by three pearls. Three additional pearls hang across the lower portion of each earring. The most popular type of jewelry in Old Russia, earrings were frequently found in treasure troves and mentioned in manuscripts, which relate that although men and women alike prized them, men customarily wore a single earring only. Many jewelers specialized exclusively in earrings, on which they lavished a wide range of materials and innovative techniques.

39 PENDANT PEARL EARRINGS

Moscow
17th century
Gold, ruby, emerald, pearl
Gold casting and engraving, enameling
Length: 4.0 cm

40 PENDANT EARRINGS

Moscow
17th century
Gold, ruby, emerald, zircon, pearl
Gold casting and engraving, enameling
Length: 6.8 cm

Shown here is a small chest known as a *larets*, used by Russians to store perfumes, jewelry, and personal keepsakes. Although made in various shapes and sizes, larets were typically rectangular lidded boxes on legs, with the four inclined panels of the lid providing a field for fanciful illustrations. The design of this chest, combining brightly colored enamel work and patterns of delicate silver filigree, exemplifies the work of the master silversmiths from Solvychegodsk. These artisans were the first in Russia to develop the technique of painting on enamel, which became widespread during the eighteenth century. On the inside lid of this box a landscape with a hunting scene provides an example of the technique.

Solvychegodsk, located at the hub of the major trade routes connecting central Russia and the Urals with the White Sea, enjoyed a booming economy in the seventeenth century as an important commercial and manufacturing center with enough surplus wealth to nurture the decorative arts. Famous toward the end of the century for their distinctive enameling techniques, Solvychegodsk's artisans perfected a style that set bright multi-colored enameled images, outlined or accented with black or dark red, against a bright white background. A profusion of flowers and plant forms, both real and imaginary, marked the designs of these masters.

41 *GOLD AND SILVER CHEST*

Solvychegodsk
late 17th century
Gold, silver, enamel
Gold gilding,
silver chasing and filigree,
enameling
Length: 19.5 cm

Painting on enamel was a specialty of the Solvychegodsk silversmiths. The combination of bright colors was accomplished by repeatedly fusing layers of paint, each of a separate color, onto the enamel base. Great skill was required to fuse each successive layer of color without melting the layer beneath.

The images on this bowl, allegorical depictions of the five senses, demonstrate the wide popularity of allegorical decoration in seventeenth-century Russian art. Floral motifs and Biblical themes are among the many devices borrowed by the Russian masters from sources in western European art, where allegory was already highly developed.

42 *SILVER AND PAINTED ENAMEL BOWL*

Solvychegodsk
late 17th century
Silver, gold, enamel
Gold gilding,
silver chasing and filigree,
enameling
Diameter: 15.8 cm

1600s

This shallow three-footed bowl, made in a traditional Russian style identified primarily with the silversmiths of sixteenth- and seventeenth-century Moscow, displays floral decorations and ornamental medallions in silver filigree with blue, white, and yellow enamel accents. The rim of the bowl is lined with a fringe of white enamel beading which appears to be suspended from threads of gold filigree. Produced in the Moscow Kremlin Workshops, as well as by craftsmen in surrounding towns, objects like this bowl illustrate a technique that was much more popular and widespread in Russia than in the countries of western Europe.

43 *GOLD AND SILVER BOWL*

Moscow
17th century
Gold, silver, enamel
Gold gilding,
silver chasing and filigree,
enameling
Diameter: 24 cm

Elaborate *panagias*, or pectoral insignias, such as this were worn on special occasions by patriarchs and other high-ranking church officials. The central design motif, an image of Christ, is engraved on a polished piece of carnelian agate and framed by gold filigree and white "pearls" of beaded enamel. The hinged rectangular clasp above the panagia carries the image of the haloed Christ, while below hangs a pendant tourmaline flanked by two sapphires.

This design continues an old tradition of setting a carved stone in the center of the jeweled panagia. Although Russian cameos began to appear in the sixteenth century, most stones predating the seventeenth century came from Byzantium or even Western Europe. Stones from local sources suitable for cameo carving were unavailable to Russian artists until the late seventeenth century, when extensive carnelian deposits were discovered in Siberia. These finds enabled Russia to mine enough stone for its own use and provide surplus for export to Europe.

44 *CARNELIAN PANAGIA*

Moscow
17th century
Gold, silver, sapphire, tourmaline, carnelian, pearl
Gold gilding, silver chasing, engraving, and filigree, enameling
Length: 15 cm

"Nikolai the Miracle Worker," the Archbishop of Mirlikiyskiy famed for performing miracles both during life and after death, was revered in the Roman Catholic church as well as the Eastern and Russian Orthodox church. Popular in Russia with both commoners and nobles, Nikolai was invoked as a protector during disasters and was the patron saint of travelers and sailors. His prestige was so high that he often appeared on icons along with Mary and Jesus.

This icon of St. Nikolai is fitted with a silver cover that frames an image of the saint painted on a wooden panel. The frame is decorated with scrolled silver-filigree designs accented with blue and white enamel and white enamel pearling – an elegant, intricate decorative scheme typical of silver and enamel objects made in Moscow during the late seventeenth century.

45 *ST. NIKOLAI ICON WITH COVER*

Moscow
late 17th century
Silver, enamel, fabric, wood, egg tempera paint
Silver chasing, engraving, gilding, and filigree, enameling
Length: 34.3 cm

This tall, graduated *stakan*, or drinking vessel, represents a type widely used in Old Russia and known from manuscript illustrations dating to the fourteenth century. The niello ornamentation on this typical seventeenth-century example, one of the finest stakans in the world, includes leafy, fruitbearing vines. Perched among the vines are large birds that look down upon a galloping unicorn, a popular image on wedding gifts.

In the seventeenth century, jewelers from Constantinople working in Moscow by special permission of the tsar strongly influenced the development of the niello technique in Russia. This exacting process, which employs a sulfurous alloy of copper, silver, and lead to obtain rich, velvety gradations of black infill, had been widely used in Old Russian arts for fine linear effects. Under Byzantine influence, Russian jewelers began to integrate Oriental motifs into their increasingly fanciful designs, adding broad fields of soft niello background to set off ornate gilded patterns of vines, leaves, fruit, and other natural forms, as in the stakan pictured here.

46 GOLD AND SILVER STAKAN

Moscow
late 17th century
Gold, silver
Gold gilding, silver chasing and engraving, niello
Height: 20 cm

Known as *stakans*, squat bowls with short, rounded legs and flat lids had been prevalent in European metal arts since the early sixteenth century. Stakans had also become popular in Russia by the late seventeenth century, although differing in proportions from their western counterparts and usually produced without lids. The central design motif on this ball-and-claw-footed example, featuring clusters of fruits, nuts, and berries intertwined with swags of gold, is borrowed from European Baroque originals.

This stakan is unique in that it bears two hallmarks. The earlier, now almost illegible, spells out the year in letters. The second stamp, a silver standard, indicates the low silver content sanctioned by government policy in 1684. The decree of Peter the Great in 1700 overturned this standard, indicating that this stakan must have been made between 1684 and 1700.

47 GOLD DRINKING VESSEL

Moscow
late 17th century
Gold, silver
Gold gilding,
silver chasing,
engraving,
niello
Height: 10.5 cm

This shallow circular plate has a wide rim decorated with a running floral pattern interspersed with scrolled leaves. The interior of the plate carries a traditional Russian design of a bird flying among intertwined vines supporting fruits, leaves, and flowers. The entire design was created using the niello technique of engraving gold images against a dark background of tarnished silver. The silver standard marking indicates a date of manufacture between 1692 and 1700.

Typical of Moscow Kremlin Workshop designs at the end of the seventeenth century, this plate's decoration reveals the influence of eastern art. Documents of the time use the term *turskaya niello*, meaning "Turkish niello," to describe the work of a Moscow master. It is not clear, however, whether the term designates a style of design or the composition of the niello alloy itself.

48 GOLD NIELLO PLATE

Moscow Kremlin Workshops
late 17th century
Gold, silver, niello alloy
Gold gilding, silver chasing and engraving, niello
Diameter: 22 cm

1600s

This large, silver-gilt, bowl-shaped *bratina*, or loving cup, is decorated with stylized views of Babylon, Persia and Macedonia rising above a scrollwork pattern of leafy vines and flanked by engravings of the Biblical Daniel and his lions. The upper rim of the bratina consists of a foliated, circular band that frames a Cyrillic inscription.

In his dreams Daniel foresaw the rise and fall of empires, as one great civilization succeeded another in an endless course through human history – hence the allegorical empires depicted here. Russia's Grand Duke Ivan III, who aggressively added territory to the state of Muscovy toward the end of the fifteenth century, eagerly seized on this notion. Dubbing Moscow the "third Rome," he inspired the rulers and officials who followed him in the next two centuries to see Russia as the vehicle of human destiny. The allegorical images on presentation pieces such as this bratina, which was used during formal ceremonies in the tsar's palace, reminded everyone present of the special historical role of the Russian state as the successor to former empires.

49 GOLD AND SILVER BRATINA

Moscow Kremlin Workshops
17th century
Gold, silver
Gold gilding, silver chasing, engraving
Diameter: 17.5 cm

The engraved gold band around the rim of this *bratina* depicts griffins interspersed with vines and floral motifs, while the sides display repeated patterns of hearts and palmettos. This piece is typical of sixteenth and seventeenth century styles, in which the artist covered simple shapes with ornate surface decoration in chasing, engraving and niello. The inscription on this bratina informs the reader that a kind person once owned the cup and that "good people should drink from it to their health."

50 *SILVER GILT BRATINA*

Moscow
17th century
Gold, silver
Gold gilding, silver chasing and engraving
Diameter: 16.2 cm

1600s

This cup, known as a *charka*, was used for tasting wine or for sampling even stronger spirits. The handle of the vessel is formed by a scroll-like vine with symmetrically placed leaves, while the interior is engraved with the image of a bird nestled among vines bearing fruit and leaves. The interior image is framed by a circular band divided into six equal sections. The charka, a popular type of drinking vessel in seventeenth- and eighteenth-century Russia, is represented by numerous examples in the collections of the Armory Museum. Most were made in gold-plated silver for a wide variety of clients or as basic retail ware.

51 *GOLD WINE CHARKA*

Moscow
late 17th–early 18th century
Gold, silver, niello alloy
Gold gilding,
silver casting, chasing,
and engraving;
niello
Length: 11.5 cm

1700s

Like the altar paten pictured on the facing page, this gold plate was made by Moscow master Ivan Christopher Yaslya as part of a full set of liturgical objects. A kneeling figure visited by a winged angel who descends from Heaven is engraved on the inner surface of the plate. High above the figure's haloed head is the dove of the Holy Spirit. The technique and composition of this image suggest a Russian master strongly influenced by exposure to Western European etchings. Inscriptions, framed by graceful leaves, fruits and flowers, border the rim of the plate.

Sixteenth- and seventeenth-century altar sets typically contained at least two or three such plates: one engraved with a Byzantine-style Madonna and Child, one carrying the image of the Crucifixion, and a third depicting John the Baptist. In the eighteenth century an Annunciation often replaced the earlier Crucifixion theme, as in this example.

1704

52

GOLD ALTAR PLATE

Moscow
1704
Gold, silver
Gold gilding, silver chasing and engraving
Diameter: 23.6 cm

This silver-gilt altar paten, a magnificent example of the art of engraving in early eighteenth-century Moscow, was made by the same master who produced the plate on the preceding page. These liturgical plates, used during the Russian Orthodox mass, held the six pieces of consecrated wafer that symbolized Heaven, the saints, Mary, Jesus, the living, and the dead. Early patens were simple flat plates; beginning in the seventeenth century, they were mounted on elaborate pedestals. The bottom was often engraved with an image of the Christ child in the manger watched over by two angels. Above the scene, a hovering dove symbolized the Holy Spirit.

53 *GOLD ALTAR PATEN*

Moscow
1704
Gold, silver
Gold gilding, silver chasing and engraving
Diameter: 27.3 cm

о санктъ · п ·
писа · андр
ѡпсопъ
· 1727

This gold snuffbox bears a miniature portrait of Peter the Great painted on copper by the master miniaturist Andrei Grigorevich Ovsov. The enamel portrait is mounted against a background of red guilloche trimmed with engraved gold swags. The Cyrillic inscription to the right of the portrait reads "In Saint P. B. Andrei Ovsov made [this] 1727." The presence of an elegant Swiss snuffbox in Russia at this time reflects Peter's aggressive adoption of Western European forms and customs; the origin of the portrait, St. Petersburg, indicates Peter's forced relocation of Russian arts, culture, and government to his newly created capital.

The portrait on the lid of the box shows the tsar dressed in armor accented by a blue sash and bearing a medal of tsarist Russia's highest order, the Order of St. Andrei Pervozvanni. Copied either from the 1717 oil portrait by Carl Moor or from an etching by Jakob Haubroken, this image of Peter is from the hand of one of Russia's first miniaturists. Ovsov was employed in the Moscow Kremlin Workshops until 1724, when he was summoned to St. Petersburg to work directly for the tsar. Only eleven Ovsov miniatures are known to have survived. The piece seen here is the only one in the collections of the Kremlin Museums.

54 *GOLD SNUFFBOX*

Snuffbox, Switzerland
1712–13
Miniature portrait, St. Petersburg
1727
Gold, copper, enamel, enamel paint
Gold engraving, enamel painting, guilloche
Length: 9.2 cm

1712

This elegant soup bowl is a superb example of the openwork filigree technique perfected by St. Petersburg's silversmiths in the early eighteenth century. The delicate, intricate filigree of silver wire creates a bold lacework that lies on the gold surface of the bowl and creates an illusion of depth on the otherwise flat surface. The shape of the bowl is reminiscent of a flower whose surface is profusely decorated by yet more leaves and blossoms.

These pieces belonged to the family of Grand Duke Dolgoruky, nobles whose varied fortunes reflected the continual political storms that rocked Russian history. Serving Russia from the seventeenth through the nineteenth centuries as leaders and statesmen, many members of the Dolgoruky line found themselves disgraced or exiled, had property confiscated, or were executed for placing the interests of their country above the whims of the tsar. This set was confiscated from Ivan Dolgoruky, who was executed at the order of Tsarina Anna Ioannovna on November 8, 1737, and added to the collection of the royal estates.

55 *GOLD TUREEN AND SAUCER*

St. Petersburg

1737
Gold, silver
Gold gilding,
silver chasing and filigree
Diameter: 15.2 cm

1737

The influence of European rococo design on Russian masters of the mid-eighteenth century is evident in the detailed high-relief chasing and engraving techniques that were combined to create this reliquary. Cast figures of the four evangelists adorn the sides of the tabernacle, and a cast image of Christ stands atop the whole. Scenes in relief of the Crucifixion, the two women at the tomb, Peter at the empty tomb, and the risen Christ appearing to Mary Magdalene complete the reliquary's figural program. Although the reliquary carries no stamp identifying its maker, the technical refinement of the work suggests that it is from the hand of a Muscovite master.

Intended to house the relics of a saint, elaborate reliquaries such as this one were often beyond the means of the ordinary Russian church. An inscription inside the piece explains that Ekaterina Saltikova, the widow of a general, sponsored this work in her husband's memory and that silver pieces from the Bogoyavlensky women's monastery in the Old Russian town of Kostroma were melted down to cast the reliquary.

1757

56 GOLD RELIQUARY TABERNACLE

Moscow
1753
Gold, silver
Gold gilding, silver casting, chasing, and engraving
Height: 69.5 cm

The designation "Our Savior Not Made by Hands," expressed by the portrait on this *panagia*, refers to a favorite Russian legend in which the image of Christ appears mysteriously on a linen cloth. The enameled portrait is framed by 485 diamonds reaching 5.5 carats in size. On the reverse of the pendant is an image of Our Lady of Vladimir, surrounded by more than 135 rubies and over one hundred diamonds. Belonging at one time to Bishop Platon of Moscow, this panagia – the most impressive in the Armory's collections – was probably presented to Tsarina Elizabeth Petrovna. The tsarina's portrait appears on the secondary diamond-and-ruby pendant suspended from the main icon. The chain holding the panagia itself contains 180 small diamonds.

57 PANAGIA: "OUR SAVIOR NOT MADE BY HANDS"

St. Petersburg
18th century
Gold, silver, diamond, ruby
Gold casting and gilding, painting on enamel
Length: 22 cm

1700s

ѡ О Н
Образъ нерукотворенны
Гда нашего Іиса Хрта.

By the mid-eighteenth century, the traditional form of the *kovsch*, or dipper, had given way to boat-shaped ceremonial pieces that were given to the heads of army units, successful tax collectors, and others in recognition of meritorious service. One such person was the tax collector Kozima Matveev, who, according to the inscription on this piece, received this kovsch from Tsarina Elizabeth Petrovna.

Eagles perched on either end of Matveev's kovsch hold laurel wreaths in their beaks and face in opposite directions, a pose that evokes the double-headed eagle on the tsar's imperial insignia. The birds' alert turn of the head, their widely spread wings, and their raised talons all signify vigilance and military readiness. This ornate presentation piece was created by Moscow's master goldsmith Ilya Gregoriev Kuchkin.

1755

58 *IMPERIAL PRESENTATION DIPPER*

Moscow
1755
Gold, silver
Gold gilding, silver casting, chasing, and engraving
Length: 40.5 cm

The lid of this impressive snuffbox carries the chased gold profile of Tsarina Elizabeth Petrovna, the daughter of Peter the Great, superimposed on a view of St. Petersburg. Beneath the tsarina's portrait, a diamond-covered eagle holds the imperial orb and staff, while to her right the sun appears with a single one-carat yellow diamond in its center. The empress gave this masterpiece of eighteenth-century goldsmithing to her favorite courtier and secret spouse, Count Razumovsky, possibly on his fiftieth birthday in 1759. The bas-relief images on the box were produced in the workshops of Jean Jorge of Paris, while the gold box itself was created by the St. Petersburg jeweler Jeremiah Pauzie (1716–1779), one of the most accomplished masters of his day.

The four side panels of the box present elaborate allegorical depictions of events that marked the early years of Elizabeth's reign. One scene depicts the role played by the elite Preobrazhensky regiment in elevating Elizabeth to the throne. A second allegory portrays the expulsion of the Braunschweig dynasty from Russia after the death of its patron, Anna Ioannovna. The third depicts the signing of a peace treaty with Sweden in 1743, while a fourth shows the apotheosis of Elizabeth, who stands on a pedestal as the universe kneels to her and offers symbols of the arts, the sciences, and literature.

A gold medal of 1735, made for Elizabeth's cousin and predecessor, Anna Ioannovna, and not visible in this photograph, is mounted in the bottom of the snuffbox.

59 *GOLD AND DIAMOND SNUFFBOX*

St. Petersburg
1759
Gold, diamond
Gold chasing and engraving
Diameter: 11 cm

1759

The silversmiths of Tobolsk, masters of the art of niello, used this exacting technique to decorate their works with glamorized scenes of daily life. The creamer, teakettle, and sugar bowl shown here are embellished in niello with hunting scenes and the figures of ladies and gentlemen engaged in polite conversation. Possibly made by master silversmith Peter Shimgin, this set was produced for Denis I. Chicherin, the governor of Siberia whose monogram and emblem adorn each piece. Chicherin, an educated man who actively encouraged the silver arts in Tobolsk, was known both as an official patron of the silversmiths and as one of their best customers.

60 *GOLD TEA SERVICE*

Tobolsk, Siberia
1774
Gold, silver, niello alloy, wood
Gold gilding, silver engraving, wood carving, niello
Height of teapot: 18 cm

This impressive presentation salver is one of only thirty-eight pieces by the famed Moscow master Alexei Ratkov that are known to have survived to the present day. The imperial double-headed eagle, decorated with ribbons and swags and encircled by a laurel wreath, appears at the center of the plate. The wide rim displays four medallions which alternate with braided ribbons and floral designs in high relief.

This salver, presented to Catherine the Great by the people of Smolensk, carries a program of complex allegorical decorations that celebrate the virtues of the monarch and the beneficence of her rule. In the center of the great imperial eagle, as though cradled in the bosom of the nation, is the municipal emblem of Smolensk: a cannon of war surmounted by a bird of peace. Surrounding the emblem is a poem extolling wisdom, kindness, and charity. Medallions bear images of Catherine, the European-educated German princess, nurturing her realm in the company of the arts and sciences. In contrast, other medallions bear the images of Russian leaders who lived in idle luxury – they wear Eastern dress, while Catherine's industrious subjects wear Western costume.

61 *GOLD AND SILVER SALVER*

Moscow
1780
Gold, silver, niello alloy
Gold gilding, silver chasing and engraving, niello
Length: 65 cm

This chalice is one of twenty pieces in the Armory Museum by Alexei Ratkov, the versatile master of both rococo and classical styles and techniques whose works have survived because they were so highly prized, collected, and preserved. Ratkov, here employing the classicism prevalent in late eighteenth-century Europe, added cast cherubim to a base decorated with laurel and acanthus leaves. The cold blue-gray enameled medallions, which complement the overall restrained ornamentation, bear portraits of Jesus, Mary, and John the Baptist. The two images on the base are framed by faceted pieces of glass which imitate diamonds – a *faux* technique widely popular at the end of the eighteenth and beginning of the nineteenth centuries.

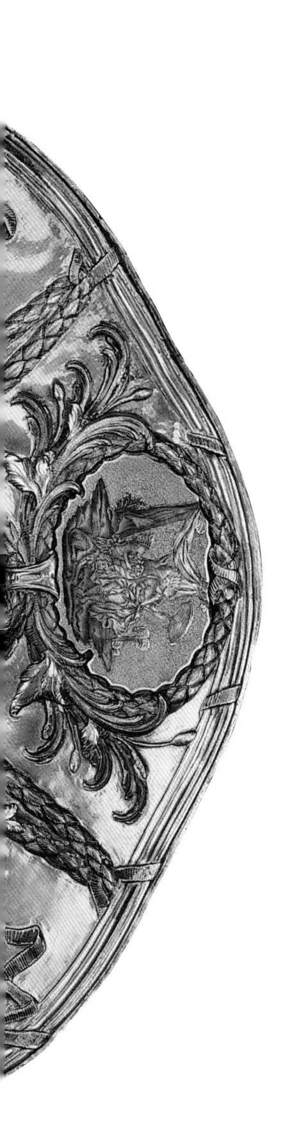

62

GOLD ALTAR CHALICE

Moscow
late 18th century
Gold, silver, glass, enamel
Gold gilding, silver chasing and engraving, enameling
Height: 43.5 cm

Mounted to the lid of this snuffbox is a locket bearing the diamond-studded monogram of Tsarina Elizabeth Petrovna, the daughter of Peter the Great who took the throne in 1742 following a palace coup. The box was created by Jean-Pierre Ador (1742–1785), who reportedly walked from Switzerland to St. Petersburg at the age of fifteen for the chance to work as a jeweler's apprentice. Ador gained high repute as a master goldsmith and became a favorite of the imperial court in St. Petersburg, where he executed commissions for the royal family from 1762 to 1785 during the reign of Catherine the Great.

Concealed inside the lid of the box is a watercolor portrait showing Elizabeth wearing an inauguration gown, an ermine mantle, and a blue sash bearing the Order of St. Andrei Pervozvanni. It is thought that this box was created to protect the portrait after the tsarina's death in 1760.

63 *GOLD SNUFFBOX OF TSARINA ELIZABETH PETROVNA*

St. Petersburg
1784
Gold, silver, diamond, glass
Gold chasing and engraving, silver casting
Length: 8.7 cm

1784

This unique gospel cover displays the skill of a consummate master of niello technique. Usually reserved for small jewelry objects or restricted to small decorative fields, niello here is used to ornament both the spine of the book and the broad expanse of the front and back panels. The intense heat required by niello, which can easily deform the shape of an object, presented a great risk to this master when he chose to decorate an entire, expensive sheet of silver.

The decorative program of this gospel, which overturns several conventions, suggests the whim of the individual who commissioned the work. On the top of the front cover the figures of Matthew, Mark, Luke, and John accompany images of the resurrected Christ, Old and New Testament interpretations of the Trinity, and the twelve apostles. The Tree of Jesse, appearing on the spine, is relocated from its usual position on the back cover. There the owner chose instead to illustrate the Crucifixion and twelve scenes from the Passion of Christ: the Last Supper, the communion prayer, the kiss of Judas, the arrest of Jesus, the presentation to Caiaphas, Peter's denial, Jesus in a purple robe, the scourging of Christ, the crown of thorns, Christ bearing the cross, the Deposition, and the Entombment. Although it is not known for which cathedral this cover was made, the extraordinary weight of book and cover – more than twenty-two pounds – suggests that it was produced for a very wealthy congregation.

64 *GOSPEL COVER*

Moscow
1794
Gold, silver, velvet, paper, wood, niello alloy
Gold gilding, silver casting and engraving, niello
Height: 53.5 cm

Gospel covers combining precious metals and gemstones were made to fit over the bound altar Bibles used in Russian Orthodox churches. The ornate floral designs on the cover seen here were produced by filigree, chasing, and granulation techniques. The hinges that connect the front and back covers to the spine allowed the massive book to lie open for reading, while two silver clasps at the front closed the book securely when it was not in use. This Bible, printed in Moscow in 1797, was purchased originally for an individual's personal use or for use in the home. The filigreed silver added later hints that the book was subsequently appropriated or donated for church use. Filigree technique saw a surge in popularity at this time, as jewelers in Moscow, St. Petersburg, and many smaller towns used it to decorate both liturgical and secular objects.

65 *SILVER GOSPEL COVER*

Russia
late 18th–early 19th century
Silver, paper, wood, niello alloy
Silver casting, filigree, and granulation; niello
Length: 14.5 cm

1800s

This elongated snuffbox, made by Otto Samuel Keibel, contains a cipher for the French phrase *amitie & reconnaissance* (friendship and gratitude). The key to the cipher is found in the first letter of each of the twenty gemstones which cross the lid from left to right. The six stones forming the word *amitie* are amethyst, malachite, "Indian topaz" (citrine quartz), topaz, Indian topaz again, and emerald. The diamond-studded ampersand is followed by the word *reconnaissance*, represented by ruby, emerald, chrysoprase, opal, nephrite (twice), agate, Indian topaz, two sapphires, amethyst, nephrite, chrysoprase, and emerald.

This sophisticated keepsake, commissioned most likely as an expensive decorative gift rather than as a functional snuffbox, reflects the fashion for French as the courtly language in eighteenth- and early-nineteenth-century St. Petersburg. It was popular among the nobility of the period to speak and correspond in French, play clever word games, and create puzzles of the sort illustrated by this box.

66 *GOLD AND GEMSTONE SNUFFBOX*

St. Petersburg
early 19th century
Gold, silver, diamond, assorted stones, enamel
Gold chasing and engraving, silver metalwork, enameling
Length: 9.5 cm

In the eighteenth and nineteenth centuries, the influence of Western European culture, especially in St. Petersburg, increased the popularity of tobacco in Russian society. Numerous jewelry firms supplied the demands of fashion by producing small, elegant, ornate snuffboxes in precious metals and gemstones that soon became popular as gifts and collector's objects. The rectangular gold box seen here was made by J.S. Barbe, a jeweler at the imperial palace and son of the famous goldsmith Carl Heinrich Barbe. The younger Barbe's career typifies the integration of foreign craftsmen into Russian society. In 1799 he emigrated from Germany to Russia, in 1806 he joined the foreign jewelers' guild, and in 1811, after becoming a Russian citizen, he entered the Russian jewelers' guild.

67 *GOLD SNUFFBOX*

St. Petersburg
1832
Gold
Gold engraving
Length: 9.1 cm

This presentation piece, created by master silversmith Sakerdon Skripitzyn, was part of a large traveling service set, so called because the sides were decorated with scenes of different cities and vacation spots. In a superb demonstration of the "smooth" niello technique to create decorative images on gold and silver, Skriptizyn adorned this dish with exquisitely detailed views of St. Petersburg. Etched around the sides of the piece is the public square in front of the imperial palace, in the center of which stands the famous Alexander Column, erected in 1834 to commemorate the defeat of Napoleon's army. A map of the Vologda region is inscribed on the cover of the dish. These once popular sets are now represented in museum collections by only two dishes, two drinking glasses, and two plates.

68

GOLD AND SILVER SERVING DISH

Vologda, Russia
1837
Gold, silver, niello alloy
Gold gilding, silver casting, niello
Diameter: 21.5 cm

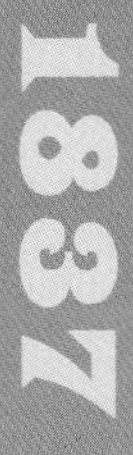

Among the nineteenth-century silversmith's favorite decorative motifs were city scenes or maps of cities, often reproduced from original etchings or adapted from original sources with new variations. Niello, offering the artist a means for rendering intricate details, was perfectly suited to this purpose. The cup and saucer shown here are decorated with views of the Moscow Kremlin as seen from the Moscow River, including the familiar, many-domed Cathedral of the Intercession, commonly known as St. Basil's Cathedral. The etched scenes are framed by the rich floral and foliate designs popularized in the sixteenth century.

69 *GOLD NIELLO CUP AND SAUCER*

Moscow
1854
Gold, silver, niello alloy
Gold gilding, silver engraving, niello
Diameter of saucer: 15 cm

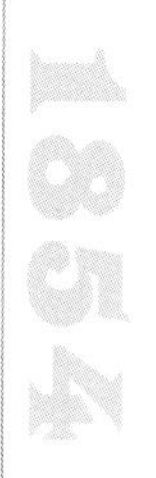

Carved from a variety of jasper often referred to as "bloodstone," this massive snuffbox combines red, green, and black colorings, a gold rococo frame, and chased rococo ornament. The outer rim and sides of the box are encased in gold scrollwork, while the lid is decorated by a floral bouquet with leaves of gold and blossoms of diamonds, emeralds, and rubies. The pedestaled vase holding the bouquet is represented by a triangular diamond weighing almost seven carats. Commissioned by an unknown owner in the mid-nineteenth century, this snuffbox was acquired by the Revolutionary Committee after 1917. An inscription in the box explains that in 1919 it was awarded by the Committee to a high-ranking officer in the Red Army named Nikolai Rattell. The Kremlin Museums acquired the snuffbox in 1982.

70 GOLD, DIAMOND, AND RUBY SNUFFBOX

Possibly St. Petersburg
mid-19th century
Gold, diamond, ruby, emerald, jasper (bloodstone)
Gold chasing and engraving
Length: 9.2 cm

This traditional eight-piece traveling ensemble, a rare original matching set with all eight pieces intact, was produced by two separate major jewelry firms. The knife, fork, and spoons were made in 1875 by I.P. Khlebnikov. On the knife handle is the figure of a man, on the fork that of a woman, and on the two spoons those of a boy and a girl respectively, all wearing traditional Russian dress. The two cups, the napkin ring, and the chair-shaped salt shaker were produced in the workshops of V. Semyonov in 1876. Their engraved niello scenes of St. Basil's Cathedral, Red Square, and the Petrovsky Palace – monuments that survived the burning of Moscow by Napoleon's troops – were favorites among artists wishing to evoke memories of traditional Russia through her architecture.

71 *GOLD AND SILVER TRAVELING SET*

Moscow
1875–76
Gold, silver, niello alloy
Gold gilding, silver casting and engraving, niello
Length of knife: 27.5 cm

1875

In a nod to the sentimental strain in late nineteenth-century pictorial arts, artists and jewelers often depicted the turtle, a symbol of tranquility and peace in the home. This well-dressed, somewhat heavy little turtle was probably worn by an older married woman. Made by the St. Petersburg master I. Frohberger, the brooch features a mother-of-pearl shell whose undulating surface imitates an actual turtle's carapace. The underside, too, is chased and engraved to replicate the texture of a real turtle shell. Two rubies comprise the eyes of this whimsical creature. Frohberger was inspired by antique brooches of the sixteenth-century Renaissance style, which often showcased large "baroque" pearls as central design elements. Here, however, he has substituted mother-of-pearl, the natural material that lines the inside of mollusk shells, for an actual pearl.

72 *PEARL AND DIAMOND TURTLE BROOCH*

St. Petersburg
1875–1900
Gold, diamond, mother-of-pearl, enamel
Gold chasing, enameling
Length: 6.7 cm

This cigarette case and matching cuff links were created by the Fabergé firm under the direction of Finnish workmaster Erik Kollin for Tsar Alexander III, whose monogram is carved on each piece. Elegant and refined, this set of gentleman's accessories illustrates Carl Fabergé's unique penchant for continuing the old Russian tradition of using common and semi-precious stones to create luxury objects. Fabergé masters and craftsmen created their own style for these objects, applying superb technical skill to boxes, vases, cigarette cases, and small carvings of animals and people.

Rhodonite, one of the most popular decorative stones in Russia, was mined from vast deposits in the Ural mountains. Many Fabergé craftsmen, themselves emigrants from the Ural region, had a natural affinity and skill for working with the varieties of stone found in the Ural, Altai, and Siberian lands of Russia.

73

RHODONITE CIGARETTE CASE AND CUFFLINKS

St. Petersburg
1881–91
Rhodonite, gold, steel
Carving, polishing
Length of cigarette case: 11.4 cm

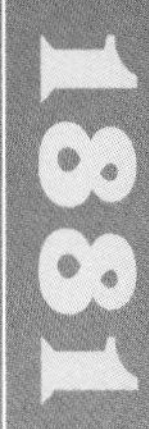

Acquired by the Kremlin Museums in 1997 on the 110th anniversary of the founding of Fabergé's Moscow division, this mother-of-pearl paper knife has special value as one of the earliest pieces created in the Moscow workshops of the famous firm. Still with its original fitted case, it bears the seal of Fabergé's Moscow branch.

The handle of the knife, a single piece of mother-of-pearl accented with floral motifs and ribbons of gold and platinum, reflects the sophisticated artistic ethic of the Fabergé firm. Somewhat out of fashion in the late nineteenth century, mother-of-pearl at this time was used primarily as an accent in jewelry, clock faces, picture frames, ladies' fans, and the handles of teapot lids. Interestingly, the Fabergé masters, like their counterparts in other firms, rarely used actual mother-of-pearl, but went to great lengths in search of processes that would simulate the look of the substance. This quest to improve on nature through artistic innovation led, for example, to "oyster enamel," a transparent white enamel with soft opalescent tones. Such creative ingenuity was an outgrowth of the Fabergé credo that the value of an object lay not in its constituent materials but in the originality and execution of its design.

The floral garland adorning the handle of the knife, executed in the *quatre couleurs* (four color) gold popularized by French neoclassicism, expresses the sentimental romanticism of the late nineteenth century. The pink rose symbolizes love, the blue forget-me-nots signify affections tested by time, and the green-gold daisies represent the innocent language of love. Metaphors for life and love were common in Fabergé products and often appeared in utilitarian objects such as this paper knife.

74 *MOTHER-OF-PEARL PAPER KNIFE*

Moscow
1890–96
Gold, silver, platinum, sapphire, mother-of-pearl
Gold casting, chasing, engraving, and carving
Length: 35.8 cm

With petals of moonstone, a golden stem, and golden stamens tipped with diamonds, this lily of the valley brooch illustrates a floral theme favored by Fabergé designers. The lily of the valley was a favorite motif at the turn of the century, a period dubbed the "Silver Century" by Russians of the time. Because poets and artists of this epoch described lilies of the valley as "pearls of moonlight," Fabergé masters illustrated the metaphor with pearls and moonstone, a semi-precious variety of feldspar.

Produced in the Moscow division opened by Fabergé in 1887, this brooch is the creation of workmaster Oskar Pihl. The student and son-in-law of the famous St. Petersburg workmaster August Holstrom, Pihl was transferred from St. Petersburg to Moscow to supervise the newly opened branch. Unlike the many other goldsmiths who worked for Fabergé, Pihl enjoyed the unique privilege of putting his own mark on the pieces he supervised. His early death in 1897 at the age of thirty-seven made Pihl pieces very rare and highly prized.

1887

75 GOLD LILY OF THE VALLEY BROOCH

Moscow
1887–97
Gold, diamond, moonstone
Gold casting and mounting, stone engraving
Length: 5.2 cm

A unique piece from the end of the nineteenth century, this brooch consists of intertwined clover leaves rendered in pear-shaped rubies and diamonds. The clover, a symbol of fertility in Old Russia, was also thought to insure the safety of the wearer. This brooch, which can also be worn as a pendant, was made in St. Petersburg by Carl Selenius, who harmonized form and color to yield a supremely elegant design.

76 RUBY AND DIAMOND BROOCH

St. Petersburg
late 19th century
Gold, diamond, ruby
Gold casting
Length: 3 cm

This massive hinged gold bracelet, marked with the initials of its unknown maker (C. S.), is decorated with applied ribbons that form regularly spaced rings. Entwined within each circular ribbon is a different kind of bird created by a combination of gold, diamonds, and one of six gemstones: emerald, sapphire, diamond, ruby, pearl, and opal.

Birds, especially domesticated species, had been used in sculpture, fresco, and mosaic since ancient times to symbolize fertility, the nesting instinct, and the protection of home and children. The birds depicted here – stork, swan, heron, crane, turtledove, and crossbill – symbolize respectively the protection of the sky and sun, fertility, inspiration, family, and prosperity.

77 *GOLD BRACELET*

St. Petersburg
late 19th century
Gold, diamond, ruby, emerald, sapphire, opal, and pearl
Gold casting and embossing
Diameter: 7 cm

This massive silver *kovsch* from Fabergé's Moscow workshop typifies the eclectic and romantic nationalist style developed by Russian jewelry designers at the turn of the century. Moving away from the strict copies of early works produced by their mid-century predecessors, these masters combined elements of European romanticism with their own reinterpretations of traditional Russian arts. In the process they completed the transformation of the utilitarian object into a purely decorative piece, as in this kovsch, whose oversize form is layered with ornate decorations that echo Old Russian originals. The handle of the vessel, for example, includes a headdress resembling a *kokoshnik* with fantastic *ryasny* pieces attached. Profuse, exuberant floral designs fit well with the grotesque form of the vase.

This kovsch also bears a brightly colored enamel illustration that both complements the fantastic form of the piece and furthers the evocation of Old Russia. This miniature is a replica of a well-known painting by V. Vasnetsov, *The Three Bogatirs*. A bogatir was a knight of Old Russia, and this subject was intended to strike a chord of nostalgic nationalism in the viewer. Vasnetsov, one of the first artists to reinterpret traditional Russian styles and motifs, influenced countless masters of the applied jewelry arts in Russia.

78 SILVER KOVSCH

Moscow
1890s
Gold, silver, quartz, smoky quartz, amethyst, chalcedony, enamel
Gold casting, chasing, and gilding, enameling
Length: 34 cm

The vibrant floral patterns decorating this matching teapot and creamer were executed in a traditional Russian style, using blue, white, and red enamels. These pieces were produced by the Moscow firm of I. Saltykov, which became famous for perfecting a variety of techniques that combined silver filigree and enamel decoration.

79 SILVER TEAPOT AND CREAMER

Moscow
1895
Gold, silver, enamel
Gold gilding, silver filigree, enameling
Height of teapot: 13 cm

1895

The unparalleled technical skill of Fabergé's master jewelers, evident in such works as this exquisite gold cigarette case, continually amazed their western counterparts. This elegant gold case with a sapphire-mounted clasp closes soundlessly on invisible hinges. When the case is closed, the seam separating its top and bottom sections is barely visible. Europe's elite prized Fabergé pieces for this superbly refined technical skill: Tsars Alexander III and Nicholas II, England's King Edward VII, the Rothschild bankers, and various industrial magnates and politicians all collected Fabergé cigarette cases. Russian jewelers of the period developed the *samorodok* style seen here, which is based on the use of rough gold surfaces that imitate the texture of natural gold nuggets. The vogue for natural-looking gold and silver articles reflected Russia's position at the time as the world's largest exporter of precious metals.

80 GOLD CIGARETTE CASE

Moscow
1899–1908
Gold, sapphire
Gold chasing
Length: 9.8 cm

Swivel-mounted on a gold handle, this Fabergé gold and topaz signet made by workmaster V. Raimer combines three facets, each with an engraved face. One side bears the emblem of Moscow. The second and third facets, respectively, display the words *Ilyinskoye* and *Usovo*, the names of two country estates near Moscow. The insignia are those of Grand Duke Sergei Alexandrovich Romanov (1857–1905), the fourth son of Tsar Alexander II. A veteran of the Russian-Turkish war of 1877–78, Grand Duke Sergei headed the elite Preobrazhensky regiment from 1881 to 1891 and in 1891 was named governor-general of Moscow. In 1905 he was killed in Moscow's Kremlin by I. P. Kolayev, a member of the Social Revolutionary Party.

81 *GOLD AND TOPAZ SIGNET*

St. Petersburg
c. 1898
Gold, topaz
Gold casting,
hardstone engraving
Width: 5 cm

1898

While Carl Fabergé's St. Petersburg shops catered to westernized, European tastes, his Moscow shops specialized in silver jewelry, tableware, and other luxury objects strongly influenced by the traditional Russian designs so popular there. This silver-gilt *bratina*, though made by Fabergé early in the twentieth century, recalls traditional Russian styles popular in the seventeenth century. In Moscow pieces such as this bratina the firm strove for a complex, archaic, decorative character at the expense of simple functionality. The surface of the vessel, covered with blue matte-finished enamel, bears images of the mythical sirens, borrowed by pagan Slavic mythology from earlier Greek sources.

82

SILVER AND ENAMEL BRATINA

Moscow
1899–1908
Gold, silver, enamel
Gold gilding, silver chasing and engraving, enameling
Height: 15.5 cm

In 1990 a cache of Fabergé jewelry, invoice numbers, and price lists was found during the renovation of an old house in Moscow. This modern treasure trove had been hidden in two metal candy boxes by V. S. Averkiev, a one-time director at Fabergé who had lived on in Moscow after the closing of the firm. In the late 1920s, Averkiev was arrested and interrogated by the KGB. Although his fate is unknown, he was apparently able to keep secret the location of the treasure that was found by construction crews sixty years later. The labels found in the candy boxes indicate the high value of the pieces, which combine large sapphires, diamonds, pearls, and rows of tiny rose-cut diamonds. The stones are mounted by means of a demanding technique that invariably raised the value of an object. Averkiev's invoices further indicated that these pieces were made during World War I, when inflation forced many Russians to invest in portable, disposable assets like jewelry, for which jewelers accordingly raised their prices.

The pieces in this set are made primarily from platinum and platinum alloys. Carl Fabergé, the first jeweler to embrace the use of platinum, recognized that this brilliant metal would neither diminish the beauty of diamonds nor conflict with their color, as gold does. The flatness and simple geometric forms of the jewelry seen here and the open mounting of the stones are characteristic of the Art Deco style, which was developed in Europe and popularized at the 1925 Paris Exhibition. Fabergé artists, designers, and workmasters were among the first to apply the new Art Deco style to jewelry.

83 *SAPPHIRE AND DIAMOND PENDANT*

Moscow
late 19th–early 20th century
Platinum, gold,
silver, diamond, sapphire
Gold filing,
platinum chain
Length: 9 cm

84 *DIAMOND AND EMERALD PENDANT*

Moscow
late 19th–early 20th century
Silver alloy, diamond, emerald
Silver casting and chasing
Width: 8.1 cm

85 *PLATINUM AND DIAMOND BROOCH*

St. Petersburg
1910
Platinum, gold, diamond, pearl
Platinum filing, gold casting
Length: 7 cm

The Fabergé firm became famous for producing hardstone carvings in semiprecious stones such as lapis lazuli, malachite, nephrite, and quartz, including several varieties of agate. Animal subjects, the most popular of the hardstone carvings produced by Fabergé, let the master carver invest considerable charm in renderings of plump pigs, proud bulldogs, noble horses, and winsome kittens. Many of these prized souvenirs were portraits of pets commissioned by their owners. The St. Petersburg industrialist E. Nobel delighted in surprising his dinner guests by concealing inside each diner's napkin a Fabergé animal intended to reflect that guest's personality. Many hardstone carvings in the Kremlin collection belonged formerly to tsarinas, since royals also found themselves charmed by these creations.

The hardstone fish seen here, unique by virtue of the multicolored carnelian employed, demonstrates how the skilled master carver could exploit the potential of the stone: the polished body of this fish creates the illusion that it is glistening wet.

86 CARNELIAN CARVING

St. Petersburg
late 19th–early 20th century
Carnelian agate
Carving, polishing
Length: 12.5 cm

This brooch, one of the most expressive pieces of Russian jewelry in the modern style, exemplifies the popularity of insect motifs with jewelers of the early twentieth century. Creating butterflies, beetles, dragonflies, and other insects challenged the jeweler to match whimsy with technical skill, as he produced realistic recreations of nature in combinations of precious metals and stones. In this dragonfly, the platinum wings are covered with rose-cut diamonds in lacy *a jour* mountings to create the illusion of transparency. The insect's body contains a single large pearl, while emeralds from the Ural Mountains form the creature's eyes and tail.

87 PLATINUM DRAGONFLY BROOCH

Russia
early 20th century
Gold, platinum, diamond, emerald, pearl
Gold casting, chasing, and filing, platinum casting
Width: 9 cm

Produced under the direction of Fabergé workmaster August Hollming, this liturgical object is a rare departure from Fabergé's more typical and familiar production of fine jewelry, *objets de fantasie*, and other secular luxury items. Made of gold, accented with pearls, and studded with four large almandines, this cross also displays Fabergé's genius with enamel by combining several different types and techniques. The background is yellow-gold guilloche enamel, the crucifix is blue guilloche enamel, and the cross is decorated with ornamental enamel over an opalescent white base. The leaves and blossoms are executed in cloisonné, the technique created, perfected, and widely used by Byzantine jewelers between the ninth and the twelfth centuries. Out of fashion since the twelfth century, cloisonné was revived by Fabergé jewelers in their endless, inspired pursuit of novel combinations of materials and techniques.

88

GOLD PECTORAL CROSS

St. Petersburg
1899–1908
Gold, almandine garnet, pearl, enamel
Gold casting and engraving, enameling
Length: 13.6 cm

Created in the Moscow workshops of P. Tereschenko, this elegant pearl- and diamond-studded brooch replicates the branch of an apple tree in full bloom, with buds and open blossoms tipped with diamonds to suggest morning dew. This large brooch actually consists of three pieces that can be worn separately or together to resemble a corsage, as pictured here. Because of the long, dark Russian winters, flowers and other symbols of spring's renewal were traditionally among the most popular motifs in Russian jewelry. Motifs such as this apple blossom were especially in vogue at the turn of the century in both Moscow and St. Petersburg.

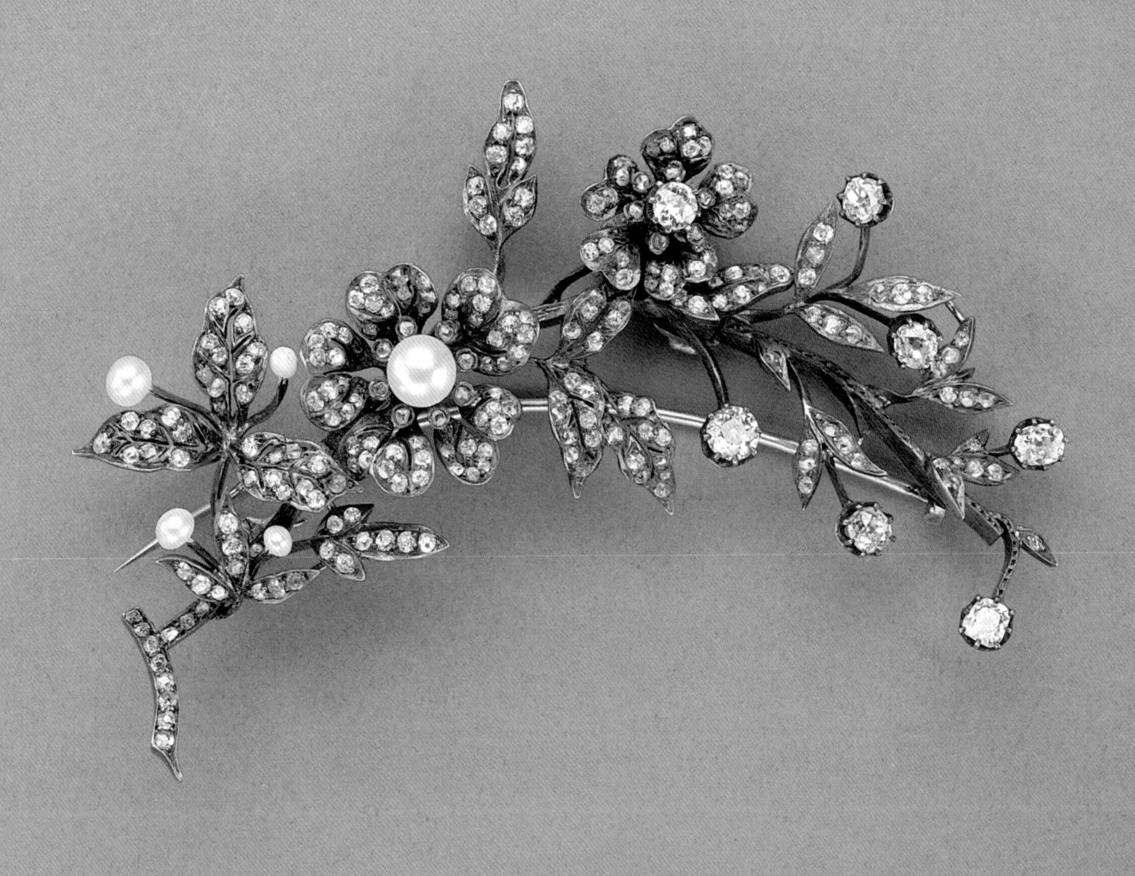

1899

89 *GOLD, SILVER, AND DIAMOND BROOCH*

Moscow
1899–1908
Gold, silver, diamond, pearl
Gold gilding, silver casting and chasing
Length: 6.4 cm

Flowers and other symbols of spring, common in Russian jewelry design in the late nineteenth and early twentieth centuries, were also favorite motifs in Art Nouveau design. The curling leaves of this iris pendant create fluid, rhythmic, organic contours that are enhanced and intensified by the effect of the colored gold frame. Rose-cut diamond accents create the impression that the leaves are tipped with morning dew. This pendant was created by Julius Rappaport, one of the most talented master jewelers to work under Carl Fabergé. Rappaport, who was born Issak Abramovich but changed his name upon conversion to Catholicism, began as an apprentice with Fabergé in 1880 and rose to the level of master in a brief four years. By the early 1900s he was the supervising silversmith of Fabergé's St. Petersburg workshop.

The rich combination of materials in this pendant exemplify the ideals of Carl Fabergé, who saw himself as an artist first and a merchant second. Continuously developing new designs in a wide variety of materials both rare and commonplace, he judged the true value of his works to lie in their artistic merit, rather than in the monetary worth of their gems and metals.

90 GOLD AND ENAMEL PENDANT

Moscow
1899–1908
Gold, diamond, enamel
Gold casting and chasing, enameling
Length: 6.2 cm

Великій Сибирскій
Желѣзный Путь
къ
1900 году.

This spectacular piece, commissioned by Tsar Nicholas II as a gift for his wife, Tsarina Alexandra Feodorovna, was one of fifty-four imperial Easter eggs produced for the royal court by the firm of Carl Fabergé. The silver egg, supported by three griffins and embellished with enamel, is engraved with a map showing the route of the Trans-Siberian Railroad, a major portion of which was completed in 1900. The egg houses a tiny model of the first train to run along this important railway, which connected the Asian and European parts of Russia. The gold train has clear quartz windows (smoky quartz in its smoking sections), ruby lanterns, and diamond headlights. Here Fabergé's trademark "surprise" – an exquisite model concealed inside the egg – is topped by yet another treat: the platinum locomotive contains a mechanism that, when wound with a tiny key, puts the train in motion.

91 *FABERGÉ EASTER EGG WITH TRANS-SIBERIAN RAILROAD MODEL*

St. Petersburg
1900
Gold, platinum, silver, ruby, onyx, quartz, enamel
Gold gilding, silver casting, platinum engraving, enameling
Height: 27.5 cm

This hollow egg, carved from a flawless piece of quartz mined in Russia's Ural Mountains, is held aloft by two lapis lazuli dolphins and embellished with lapis lazuli eagles, pendant pearls, rose-cut diamonds, and gold trim. The egg holds a tiny, exquisitely detailed gold and platinum model of the imperial yacht "Standart," the favorite vessel of Tsar Nicholas II, who commissioned this piece from the firm of Carl Fabergé as a present for his wife, Tsarina Alexandra Feodorovna. The actual yacht, nicknamed "the floating palace," was one of the great technological achievements of its time: 160 meters long and lavishly appointed, with accommodations for the royal family and more than 300 servants and crew members.

92 FABERGÉ EASTER EGG WITH MODEL OF THE IMPERIAL YACHT "STANDART"

St. Petersburg
1909
Gold, silver, quartz, lapis lazuli, pearl, enamel
Gold chasing, silver engraving, enameling
Height: 15.3 cm

1909

This bulbous silver and enamel *kovsch* was made for the Fabergé firm by Feodor Ruckert, the silversmith who produced almost all of the cloisonné enamel pieces sold by Fabergé. In addition to filling orders for Fabergé, Ruckert also supplied several of the firm's competitors, including the prominent Ovchinnikov workshop. Working almost exclusively in the Old Russian style, Ruckert created combinations of delicately balanced complementary colors, rather than the bold contrasts of color favored by his predecessors.

93 *SILVER AND ENAMEL KOVSCH*

Moscow
1908–17
Silver, enamel
Gold gilding, silver filigree and chasing, enameling
Length: 15.5 cm

The diamond-studded wings of this ruby-eyed stag beetle swing open at the touch of a button to reveal a timepiece mounted in the beetle's body. The shimmering effect on the surface of the wings was achieved by applying translucent enamel over cross-hatched patterns engraved in the gold. The watch, made in Switzerland, carries the mark of the Riga District Assay Office import division. This swift-flying beetle, a witty allusion to the adage that "time flies," reminded the wearer of the accelerated pace of life in the twentieth century.

94 GOLD PENDANT BEETLE WATCH

Timepiece, Switzerland
1908–17
Beetle, Moscow
1908–17
Gold, diamond, ruby, glass, enamel
Gold casting and engraving, enameling
Length: 5.7 cm

Expensive, naturalistic brooches made in the form of New World beetles first appeared at the World's Fair Exhibition in Paris in 1867 and marked an original, innovative departure from the prevalent Greek-, Egyptian-, and Renaissance-inspired jewelry styles of the time. Catching on in Russia, beetle brooches made in a variety of materials and combinations of contrasting stones were widely popular for several decades. The beetle seen here was made in a severe, refined style, its strong simple colors achieved by combining cold blue sapphires, snow white mother-of-pearl, and small sparkling diamonds.

95 *MOTHER-OF-PEARL BEETLE BROOCH*

St. Petersburg
1908–17
Gold, diamond, sapphire, mother-of-pearl, glass
Gold and silver casting
Length: 6.5 cm

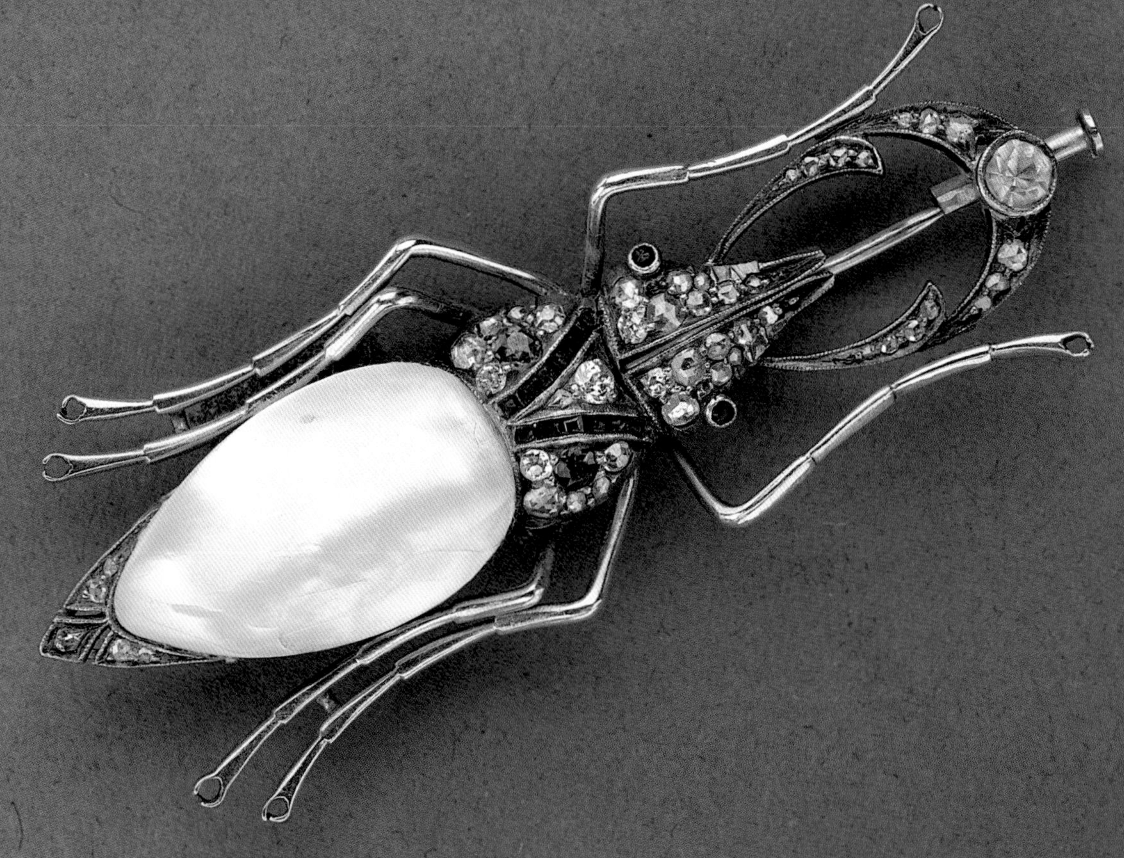

Like the snuffboxes of the previous century, elegant cigarette cases enjoyed wide popularity among the Russian nobility during the late nineteenth and early twentieth centuries. Prominent jewelry workshops, among them Fabergé, F. Kehli, K. Blank, and I. Britzin, all produced superb examples. This case was made by Britzin, whose works were of such high quality that he was long thought to have worked for the Fabergé shops. The son of a farmer who lived near Moscow, Britzin moved to St. Petersburg at the turn of the century to open a shop specializing in small luxury articles such as clocks, cigarette cases, and tableware. Britzin cigarette cases rivaled those of Fabergé in beauty and technical refinement. This one closes quietly on invisible hinges, has surfaces that are polished to a mirror finish, and features soft-colored enamels of an opalescent sheen that was achieved by polishing successive layers of enamel with a special wooden wheel.

96 GOLD AND ENAMEL CIGARETTE CASE

St. Petersburg
1908–17
Gold, silver, enamel
Gold gilding, silver engraving, enameling
Length: 8.5 cm

The geometric and floral designs on this silver sugar bowl and creamer are reminiscent of Old Russian ornament, with patterns outlined in silver filigree and selectively filled in with multicolored enamels. These traditional designs were produced by artisans in Moscow, who, by the early twentieth century, had begun to organize themselves in guild-like organizations called *artels*. By 1916 Moscow alone had more than thirty such artels. The pieces shown here were produced by the silversmiths and enamelers of the Odinnadtsaya (11th) artel.

97 *SILVER SUGAR BOWL AND CREAMER*

Moscow
1908–17
Gold, silver, enamel
Gold gilding,
silver filigree,
enameling
Height of creamer: 8.5 cm

1908

Elegant accessories such as this little purse expressed both the aesthetic ideal of the Art Moderne era and the conviction that each element of the stylish woman's outfit – purse, lorgnette, bag, fan, boa, or stockings – was equally important in creating an overall impression. The chain mail mesh of this purse is composed of strands of gold wire so fine that they achieve the suppleness of actual fabric. The decorations of the frame alternate round diamonds with square-cut rubies. A cabochon ruby adorns the clasp.

This purse bears the mark of the St. Petersburg District Assay Office, which sets the purity of the gold used at fifty-six *zolotnicks* – approximately equal to fourteen karat in the Western system of measures. In the Russian system, ninety zolotnicks was considered to be pure gold and equaled the Western standard of twenty-four karat. The design of the assay mark also helps to establish a firm date for the manufacture of this piece.

98 *GOLD AND SILVER MESH PURSE*

St. Petersburg
1908–17
Gold, silver, ruby, chamois leather
Gold casting and engraving
Length: 16.5 cm

This punch bowl, made for the St. Petersburg retail firm of Viktor Gordon, combines a silver lid, handles, and dipper produced by S. Pavlov.

Punch bowls, borrowed from English culture, were popular in Russia for serving punches made from such wines as muscatel and Hungarian Tokay. Cut crystal vessels such as this one helped set a refined, elegant tone at exclusive dinner parties attended by dashing officers from elite military corps. The complex diamond pattern cut into this tall, cylindrical vase would have enhanced the glamour of such a soiree by giving off a beautiful glow of color as it refracted the deep hues of the wine punch within.

The cover of the bowl is expressive of nineteenth-century Russian romanticism, with its love of poetry and fairy tales. It bears an image of a "sleeping head" taken from Pushkin's romance *Ruslan and Liudmila*. Offsetting this almost lurid image is the neoclassical detailing of the silver rim and handles.

99 *SILVER PUNCH BOWL*

St. Petersburg
1908–17
Silver, gold, lead crystal
Silver casting, chasing, and engraving; gold gilding
Height: 46 cm

An early twentieth-century Russian nostalgia for a fabled romantic past is expressed vividly in this inkpot. Made of glass and joined to a marble base by a silver mount in the shape of a tree stump, the inkpot includes the figure of an Old Russian knight. Russian folklore – especially themes involving knights and other fabled protectors of the ancient Russian homeland – was frequently combined with a neo-Russian Moderne style in the decorative arts of this time. The preeminent workshop of I. Khlebnikov, for example, exploited this theme often.

100 SILVER INK POT

Moscow
1908–17
Silver, glass, marble
Silver casting, chasing, and engraving
Height: 26 cm

1908

An eclectic mix of styles, typical of Russian decorative arts in the early twentieth century, marks this silver and crystal dish from the leading Moscow silversmithing firm of I. Khlebnikov. An arctic theme is expressed by the polar bears standing atop icicle-laden ledges and by the silver ice crystals that line the rim of the Russian crystal dish, while the form of a traditional Russian *kovsch* is suggested by the tall handle. Besides combining the animal style and the neo-Russian style, this piece also partakes of the Art Moderne influence with its smooth surfaces and absence of hard lines.

101 *SILVER AND CRYSTAL "ARCTIC" DISH*

Moscow
1908–17
Silver, lead crystal
Silver casting and chasing, crystal cutting
Length: 39 cm

This low-relief, leaf-shaped crystal candy dish was produced by the Nechayev-Maltzev factory, a workshop near Moscow that is still in operation, and bears the Gus Khrustalny trademark. The handle, in the form of a large stylized bird, was added by the firm of I. Khlebnikov, the preeminent Moscow silversmiths who frequently received commissions to embellish Russian crystal. Khlebnikov, the foremost Russian practitioner of the Moderne style, here introduced a grotesque, bald-headed African bird, the marabou stork, which perches on the edge of a crystal leaf. A boa of multi-colored marabou feathers was highly prized among the fashionable women of Russian high society.

1908

102 *SILVER AND CRYSTAL CANDY DISH*

Moscow
1908–17
Silver, lead crystal
Silver casting and chasing, crystal cutting
Length: 29.6 cm

These matched crystal decanters, with handles and rims fashioned in the form of striking cobras, were produced through the collaborative efforts of two Moscow firms. The crystal was made in the Nechayev-Maltzev factory under the Gus Khrustalny trademark, while the silver mountings were produced by the O. Kurliukov workshops. Like the "arctic" dish pictured in plate 100, this ensemble combines an unbridled animal element and classic Art Moderne restraint: the aggressive natural forms of the cobras are balanced by the smooth simplicity of the crystal vase.

103 PAIR OF SILVER AND CRYSTAL DECANTERS

Moscow
1908–17
Silver, lead crystal
Silver casting, chasing, and gilding
Height: 30.5 cm

Illustrating a practice common among producers of Russian decorative arts at the turn of the century, these decanters combine cut crystal imported from Western Europe and silver mountings created in Russia. Here the silver pedestals, handles, and lids were made by the exclusive private firm of Nemirov-Kolodkin. This pair of elegant decanters offers an outstanding example of their work.

104

PAIR OF SILVER AND CRYSTAL DECANTERS

Moscow
1908–17
Silver, gold, lead crystal
Silver casting and chasing, gold gilding, crystal cutting
Height: 37 cm

1908

This elegant table piece, a crystal vessel imported from Western Europe with a silver rim and handles produced by Moscow's Fourth Workman's Artel, was used to chill bottles of wine on formal occasions. The large size of the piece, the massive crystal, and the richness of the decoration suggest that it once belonged to a set of royal tableware used at the most formal of dinners and receptions. The master who created it achieved such a fine proportional balance between the silver work and the crystal as to make this very heavy piece appear almost weightless – a feat of great subtlety and skill.

105 *SILVER MOUNTED CRYSTAL TABLE PIECE*

Moscow
1908–17
Silver, lead crystal
Silver casting, chasing, and gilding; crystal cutting
Height: 26 cm

Made by Fabergé workmaster Henrik Wigstrom, this gold-framed enameled notebook cover was acquired by the Kremlin Museums in 1991. As a *carnet du bal* this notebook would have been used by a fashionable lady at a Russian ball to note the names of each dance and dance partner who graced her evening. An example of the most popular type of the highly detailed, exquisitely crafted luxury items produced by Fabergé, this cover features transparent enamel colors laid over an engraved guilloche ground. Fabergé craftsmen were the world's unequalled masters of enameling, capable of producing over 144 different color tones. Some of these colors became Fabergé signatures that no other firms could reproduce, including oyster white, sturgeon pink, and royal blue. In this cover a light blue enamel overlays engraved guilloche patterns that resemble rays of sun. At the center, the opalescent enamel displays the motif of a floral basket framed by chased floral garlands of four-color gold in the Louis XVI style.

106

GOLD AND ENAMEL NOTEBOOK COVER

St. Petersburg
1908–17
Platinum, gold, pearl, leather, enamel
Gold casting and chasing, enameling
Length: 8.5 cm

1908

This bracelet, brooch, and ring were produced by the Russian Jewels Manufacturing Corporation of Leningrad and illustrate the character of jewelry design in the late Soviet period. Composed of Russian diamonds and gold from remote Siberian mines that remain active today, all three pieces were designed by Valentina Ignatieva. Ignatieva, a star among the first generation of Soviet jewelry designers, was a restless innovator who continually sought combinations of new designs, new styles, and new materials. Her boldly expressive designs brought her the Grand Prix at an international exhibition in Brussels. The pieces seen here employ mounts that are cast to resemble starbursts – a fitting complement to the sparkle of the large, individually mounted diamonds.

107 *GOLD AND DIAMOND BROOCH*

Leningrad, USSR
1975
Gold, diamond
Gold casting, chasing, and polishing
Diameter: 4.3 cm

108 *GOLD AND DIAMOND BRACELET AND RING*

Leningrad, USSR
1975
Gold, diamond
Gold casting, chasing, and polishing
Length of bracelet 19 cm
Length of ring: 1.9 cm

This brooch belongs to a class of jewelry produced for export during the Soviet period by the state-owned Moscow Experimental Jewelry Factory. Typically, these pieces were produced by hand and relied little on the industrial technology available to the modern jeweler-craftsman. The brooch pictured here was produced by A. Petrova, one of a group of talented designers who flourished from the early 1960s through the 1980s. Like many of his forebears of the early twentieth century, Petrova creatively combined various styles. This golden knot brooch, for example, mixes an historical classicizing style with fluid Moderne elements.

Designed by B. Gladkov and created by M. Lesik, this necklace of floral and clover motifs is intended to remind the wearer of the beauty of springtime – a favorite theme born of the harsh Russian winters. Composed of gold, diamonds, and emeralds mined from within the borders of the Soviet Union, it was made exclusively by hand as a one-of-a-kind object never intended for series production. Gladkov and Lesik were among the first generation of masters who redeveloped Russia's jewelry industry, which had been dormant for much of the twentieth century. This piece, with its distinctive interpretation of floral motifs, typifies the designs of the Ekaterinburg school.

109 *GOLDEN KNOT BROOCH*

Moscow Experimental Jewelry Factory, USSR
1975
Gold, diamond
Gold casting and mounting
Width: 5 cm

1975

110 *GOLD AND DIAMOND "SPRINGTIME" NECKLACE*

Sverdlovsk, USSR
1978
Gold, diamond, emerald
Gold casting, chasing, and engraving
Length: 19 cm

1978

This brooch was designed in 1980 by T. E. Vasiliyeva at Moscow's Experimental Jewelry Factory and executed in gold, diamonds, and enamel by I. Grakovsky and V. N. Kolobulin. Made exclusively by hand and not intended for mass production, the Lubava Brooch was one of the first pieces to demonstrate the redevelopment of the enamel technique that had been so widely popular in the early decades of the twentieth century.

This exclusive jewelry ensemble, known as the "Snake" set, is a unique edition never intended for mass production. Designed by Lubov Shepel, chief designer at the Kiev Yuvelirprom Company, this magnificent set won a special citation during an international jewelry competition sponsored by the DeBeers company.

111 *LUBAVA BROOCH*

Moscow Experimental Jewelry Factory
1980
Gold, diamond, enamel
Gold casting, chasing, and engraving
Length: 5.1 cm

1980

112 *GOLD NECKLACE, BRACELET, EARRING SET*

Kiev Yuvelirprom
1985
Gold, diamond
Casting, engraving, polishing
Necklace 39.5 cm
Bracelet 22 cm
Earrings: 4 cm

1985

Made for Russia's chess grandmasters Anatoly Karpov and Gary Kasparov to commemorate their famous match of 1987, these decorative intaglio chess queens were made between 1986 and 1988 by P. Potekhin. Potekhin was a leading designer at the St. Petersburg School of Applied Arts, the institution founded in the nineteenth century by Baron Stieglitz and renamed in 1968 for V. I. Muhina. Potekhin was one of the first jewelers to revive intaglio carving in Russia. These chess queens are the two largest Russian intaglios ever created.

1986

113 *SMOKY QUARTZ INTAGLIO CHESS QUEENS*

Leningrad
1986–88
Smoky quartz
Intaglio carving
Height: 25.9 cm

This dazzling presentation crown was created for the 500th anniversary of the founding of the Annunciation Cathedral in the Moscow Kremlin. Its primary motif is a stylized *kokoshnik*, the traditional onion dome that graced typical Russian Orthodox churches. The kokoshnik, symbolic of Russia's ancient past, is intended to transport the viewer from the end of the twentieth century to the early epoch of traditional Russian architecture. Covered with Yakutia diamonds from Siberian Russia, decorated with large faceted amethysts, and surmounted by large pearls, this creation recalls the regal opulence of the crown produced for Tsar Ivan Alexeivich pictured in plate 35. The crown is the work of designer Valery Golovanov, who was born in 1957 and graduated from Moscow's School of Arts before entering the Moscow Experimental Jewelry Factory. Since 1992, Golovanov has worked for the Parura Company, which provides Russian museums with museum-quality replicas and restorations of historic pieces.

114 *DIAMOND AND AMETHYST CROWN*

Moscow Experimental Jewelry Factory
1990
Gold, diamond, amethyst, pearl
Gold casting and mounting, montage
Height: 14.7 cm

This box was created by M. Maslenikov, a member of the Union of Artists of the Soviet Union whose works are strongly influenced by the development of a Russian *avant-garde* mosaic style. Unlike the craftsmen of the Russian State Jewelry Factory, who enjoyed the use of the finest precious metals and stones, jewelers in the Union of Artists typically worked with common stones, metal alloys, glass, wood, and other "non-jewelry" materials. The *avant-garde* style they pioneered, however, never denied the many centuries of artistic tradition and technique that preceded them. This box, for example, includes motifs taken from the late eighteenth and early nineteenth centuries and reflects the Chinese styles popular at that time. The traditional form of the clasp, with its delicate white mother-of-pearl flowers inlaid in a black mother-of-pearl ground, demonstrates the inspiration that this contemporary master found in Russia's rich artistic heritage.

115 MOSAIC INLAID BOX

Moscow
1992
Gold, silver, emerald, mother-of-pearl, coral, malachite
Mosaic inlay and chasing
Width: 8.4 cm

1992

[illegible] Chavushian's contemporary Easter egg, dedicated to Russia's last tsar, Nicholas II, draws inspiration from the renowned imperial eggs created by Carl Fabergé. A traditional symbol of life itself, and also signifying Russia's memory of her rich history, the egg employs the old Russian technique of filigree to suggest the continuous thread binding jewelers of the present to their forebears in the past. Chavushian, a graduate of Moscow's Artistic Industrial School, has exhibited widely in domestic and international exhibitions and since 1993 has been a member of the International Federation of Artists. In addition to pieces owned by the Kremlin Museums, his work is represented in the State Historical Museum in Moscow and in various collections internationally.

1993

116 *"GOLDEN THREAD" EASTER EGG*

Moscow
1993
Gold, silver, demantoid garnet, opal, chrome diopside
Filigree
Height: 14 cm

This cameo by Alexei Dolgov, in a frame made by Boris Sokolov, presents a contemporary variation on a familiar traditional format. Recalling eighteenth-century cameos of royal or noble individuals, this modern version commemorates the coronation of Elizabeth Petrovna. The empress rides in a carriage in the foreground; behind her Moscow's Assumption Cathedral can be seen in the distance. These contemporary artists have introduced dynamic elements into the traditional static cameo format, such as the fluid, organic angles of the mounting and construction that create multiple viewing perspectives. The juxtaposition of the carriage to the wheel lends a sense of movement to the composition.

117 *CAMEO PLAQUE: "DAUGHTER OF PETER THE GREAT"*

St. Petersburg
1994–95
Shell, silver
Carving
Length of frame: 12 cm

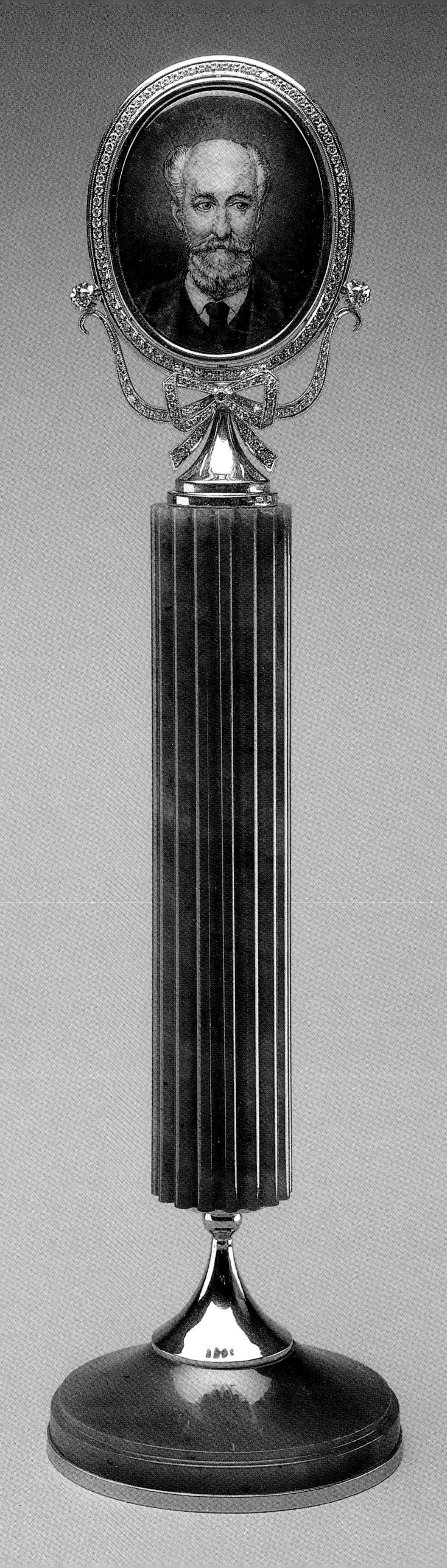

This commemorative stele was created to mark the 150th anniversary of the birth of Carl Fabergé, the genius of artistic innovation who brought international acclaim to Russian decorative arts in the early twentieth century and who continues to inspire Russian jewelry designers today. The stele was made in the Samotsvety workshop, one of the oldest and largest jewelry factories in Russia, which has long been honored with prestigious international awards at exhibitions in Russia, in the United States, and throughout Europe. The classical style of the piece reflects the traditional approach adopted by St. Petersburg's designers both in previous centuries and today. In addition, like many Fabergé works themselves, this stele is the product of a creative collaboration: conceived and designed by Svetlana Berezovskaya, a leading designer at the firm for over thirty years, this piece incorporates the skills of painter Y. Novikova, carver V. Yurkevich, and jeweler S. Bogazhenkov.

1996

118 STELE WITH PORTRAIT OF CARL FABERGÉ

Russian Samotsvety, St. Petersburg
1996
Gold, diamond, nephrite, enamel
Carving, enamel painting
Height: 26.5 cm

This elegant flower brooch was produced by the jewelry firm of Bratina Jewelers, which was founded in 1990 to specialize in exclusive, one-of-a-kind jewelry creations. Merging historical romanticism, classicism, and elements of Art Moderne style in their works, the masters at Bratina have revived traditional techniques of working with precious metals, gemstones, and mother-of-pearl.

119 FLOWER BROOCH

Moscow
1997
Gold, diamond, green tourmaline, mother-of-pearl
Gold casting, gem carving
Diameter: 5 cm

MINERALOGY

PRECIOUS STONES AND NOBLE METALS: HISTORIC SOURCES FOR RUSSIAN GEMS AND JEWELS

Joel A. Bartsch, Curator of Gems and Minerals, Houston Museum of Natural Science

Regardless of where political boundary lines have been drawn over the past millennium, Russia and the territories that make up the Confederation of Independent States have always held immense deposits of gem crystals, precious metals, and decorative stones. Those deposits have provided the raw materials for artists and jewelers, who used them to produce a staggering array of beautiful objects. One sixteenth-century visitor exclaimed that "in the Tsar's palace in Moscow there is so much gold and silver that it is almost impossible to count all the vessels."

Throughout Russian history, gemstones have been prized by the land's diverse cultures, by every segment of society. The earliest civilizations treasured gems not only for their beauty but also for their supposed magical powers. Such beliefs continued for centuries; Russian folklore is rife with allusions to the mystical qualities of gems. Diamonds had the power to mitigate lust, emeralds could restore wisdom, and sapphires would ensure tranquillity. Ivan the Terrible (1530–1584) believed the red *yakhont*, as ruby was then known, imparted a healing effect on the "strength and memory" of man.

In the sixteenth century, gemstones, gold, and silver were used as displays of wealth by members of the Imperial Court and by the bishops of the Russian Orthodox church. In 1589, Bishop Arseniy of Elasso described with wonder an official reception in the Golden Chamber of the Tsarina:

It was impossible to look at the Tsarina without being amazed at her magnificently adorned attire. On her

head she had a dazzling crown, skillfully made up of gemstones and pearls, that was divided into twelve equal sections. There were numerous carbuncles [red stones, commonly garnets], diamonds, topazes, and round pearls on the crown and large amethysts and sapphires studded around it. In addition on both sides fell triple long chains which consisted of extremely valuable gemstones and were covered with glittering round emeralds so large that they were simply priceless. The smallest part of this decoration would be sufficient to adorn ten sovereigns!

This obsession with gems is particularly remarkable in light of the fact that there were no mines operating in Russia until the early eighteenth century. In order to satisfy the demand for rare and beautiful gemstones, as well as for gold and silver, the tsars had agents around the world purchase gems on their behalf. In the late fifteenth century, Ivan III had agents permanently stationed at ports in the Crimea, specifically to acquire *lals* (zircons), *yakhonts* (rubies and sapphires), and large pearl kernels for him.

Throughout the sixteenth and seventeenth centuries, the tsars stocked the Imperial Treasury with gems acquired from traders in Genoa and Venice, as well as from merchants in Asia, Western Europe, Byzantium, and Greece. As early as the thirteenth century, gems flowed to Russia from the mines of Ceylon (now Sri Lanka), which produced rubies, garnets, amethysts, topazes, and blue sapphires of the finest quality. India and Burma (now Myanmar) provided rubies, garnets, tourmalines, jasper, agate, and carnelian. In the fifteenth century turquoise was mined in northern Tadzhikistan, although the finest quality stones were imported from Persia (now Iran). Excellent examples of this light blue Persian turquoise can be seen in the fifteenth-century gold crown made to adorn the "Mother of God of Bogolubskaya" icon [*plate 15*]. Before the sixteenth century, emeralds were brought to Russia from Egypt; later, from the mines of Muzo in present-day Colombia.

At the end of the seventeenth century, Peter the Great (1672–1725) began to actively promote the search for gems within Russian territories. Success came first in the Ural Mountains, where several gem minerals, including "rock crystal" quartz, smoky quartz, topaz, and beryl were found in river gravel. In 1723, mining operations began at Adun-Chilon when aquamarine crystals were discovered in vein deposits that outcropped at the surface.

Bolstered by those early successes, Catherine the Great continued the quest for domestic sources of gemstones when in 1765 she appointed General Yakov Dannenberg to be commander of an expedition to search for "diverse colored stones" in the area around Ekaterinburg. As a result of that expedition the famous amethyst deposits at Talyan (from "Italian," a reference to the Italian miners hired to work the deposit) were discovered. By 1770 the mining operations in that district were producing amethyst from more than one hundred separate veins.

Because the Ural Mountains proved to be a treasure trove of gemstones, the Dannenberg expedition lasted twenty-five years. When completed, more than thirty varieties of gemstones had been presented to Catherine, who used the crystal clusters "to adorn the palace." Over the next two centuries Russia's leaders continued to send expeditions to every corner of their territory in search of precious stones and industrial minerals. Their efforts led to the discovery of immense deposits of gold, platinum, diamonds, and emeralds as well as many other semi-precious stones and decorative minerals. These extraordinary native riches have enabled Russian artisans, from the earliest days to the present, to perfect the exquisite craftsmanship so visible in the artifacts presented in the exhibition *Kremlin Gold.*

NOTE: CITIES, TOWNS, AND VILLAGES MENTIONED IN THE TEXT THAT FOLLOWS ARE INDICATED BY A POINT ON THE MAPS. REGIONS OR GENERAL FEATURES, SUCH AS MOUNTAIN RANGES, ARE INDICATED BY LARGE TYPE.

NATIVE GOLD

For thousands of years gold has been treasured by civilizations the world over as the most precious of all metals. Evidence of this esteem is found in the enormous efforts expended by early goldsmiths to create intricate objects of adornment. A superb example of this high level of craftsmanship is a flexible braided gold bracelet [*plate 1*] made in the northern Black Sea region in the fourth century AD. Although gold had been highly prized here for more than a thousand years, the gold used to produce these objects was imported; there were no domestic sources of the precious metal.

This situation changed dramatically in the mid-eighteenth century, when gold ore was finally discovered in Russian territory. In 1745 a peasant named Raskolnik Markov discovered "curious yellow stones" in the Berezovsk District of the southern Ural Mountains. His discovery quickly blossomed into a bonanza that greatly increased the nation's wealth. Between 1745 and 1806 the Imperial Treasury was enriched by the addition of more than 2,800 kilograms of gold from the mines at Berezovsk.

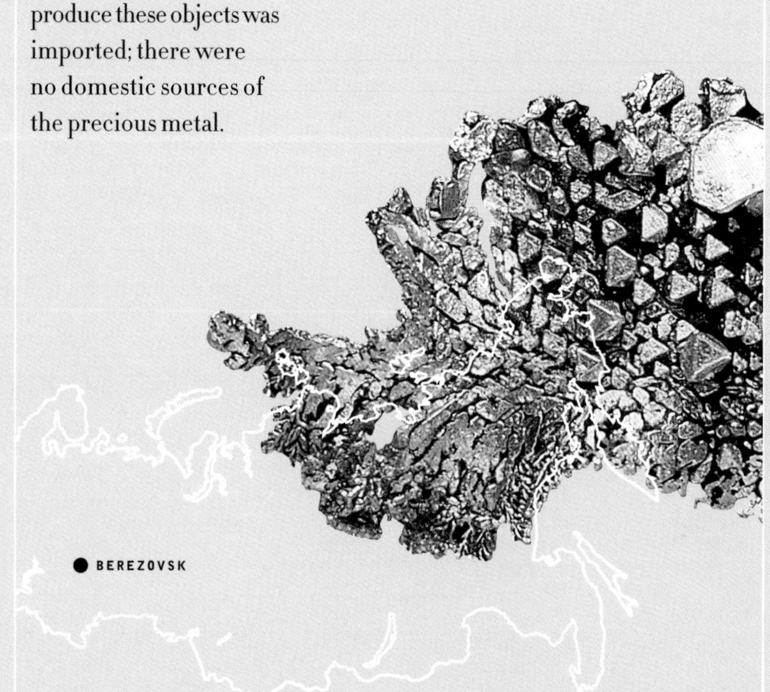

GOLD NUGGETS

Shortly after hard-rock mining commenced at Berezovsk, the director of mining operations began searching nearby streams and rivers for gold nuggets, which, he reasoned, had been washing out of the surrounding mountains and into nearby streambeds for millions of years. His assumption proved correct; over the course of the next 250 years, millions of ounces of gold nuggets were recovered from such "placer"-type (i.e. stream-borne) deposits throughout Russia.

In the early nineteenth century, the largest Russian gold nugget ever recorded was found at the Tsarevo-Alexandrovsky workings in the Berezovsk District. Called the "Zolotoi Treugolnik" (Golden Triangle) because of its shape, the nugget weighs more than 36 kilograms. In 1848 it was transferred to St. Petersburg, where it remains today. In this century, dozens of nuggets weighing more than ten kilograms have been recovered from placer mines in Siberia, including a 12.2 kilogram nugget found at Aprelskiy near the Bodaibo River in November 1990.

BEREZOVSK

NATIVE SILVER

Silver has been used by artisans in Eastern Europe for more than a thousand years to create intricately detailed objects. Early examples of the high level of skill attained by craftsmen in the region are the silver *kolts*, or medallions [*plate 8*], found in an ancient Russian treasure trove. Dating to the twelfth century, a single medallion in this trove contains thousands of individual silver beads, each surrounded by a filigreed silver band. Beginning in the fourteenth century, "golden" objects were made in silver and coated with a thin layer of gold. The gilt surfaces of these objects reflected the beauty of pure gold without the equivalent expense.

Silver was acquired through trade from mines in Saxony as early as the fifteenth century and from mines in Norway beginning in the early 1620s. In Russia, small-scale silver production began in 1707, when a smelter was constructed at Nerchinsk to extract the metal from low-grade ore found in the region. In 1732 silver was discovered in veins of white calcite on Medvezhiy Island in the White Sea. Those mines were producing enough silver by 1736 that the Tsarina Anna Ioannovna ordered that all rubles minted that year should not only bear her likeness but also be made exclusively from Russian silver. Any excess silver was reserved for the metalsmiths of the imperial court. Unfortunately, the Medvezhiy Island mines were exhausted by the middle of the 18th century; it was nearly 60 years before Russia resumed domestic silver production, when discoveries were made at Zmeinogorsky in the western Altai Mountains. In recent years, mining operations at the Sarbay deposit in Kazakhstan have encountered cavities in solid rock containing large crystallized wires of native silver.

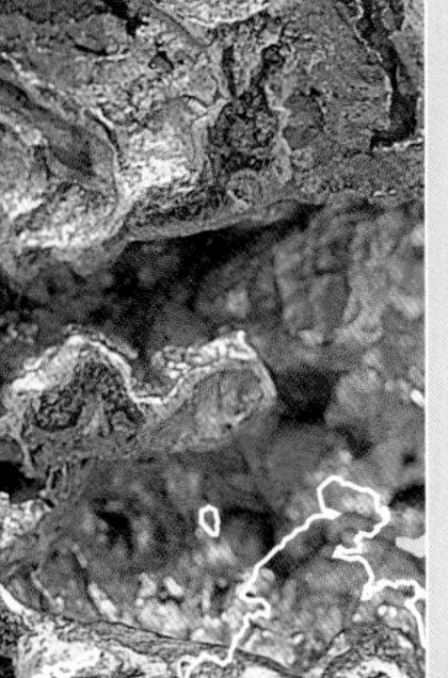

BODAIBO

ZMEINOGORSKY

NATIVE PLATINUM

Platinum, an exceedingly rare precious metal, was first discovered in the Ural Mountains at Verkh-Neivinsk in 1822 as pea-sized, stream-worn nuggets in the gravel of the Tura River and its main tributary, the Iss River. Since then, platinum has been found throughout Russia as microscopic grains in rock, as cubic crystals measuring up to two centimeters on any edge, and as nuggets weighing, in some cases, more than ten kilograms. (In contrast, the largest platinum nugget ever found outside of Russia weighed only 34 grams.)

One of the most prolific districts in Russia for the production of platinum is the Nizhny-Tagilsk district in the Perm region, which extends from Denezhkin-Kamen, north of Bogolovsk, to south of Ekaterinberg along the eastern edge of the Ural Mountains. Platinum ore has also been produced from mines in Siberia at Talnakh and Norilsk, north of Krasnoyarsk, and from the Konder massif in Khabarovsk. These mining operations have made Russia one of the leading producers of platinum, which today is used mainly for industrial applications. Although its high melting point made it difficult to work with, platinum became a popular metal for use in jewelry in the late nineteenth and early twentieth centuries. The Dragonfly Brooch [*plate 87*] made in Moscow at the turn of the century and the Diamond and Pearl Brooch [*plate 85*] made in St. Petersburg in 1910 are two beautiful examples of early twentieth-century platinum jewelry.

DIAMOND

Diamonds (crystallized carbon) were first discovered in stream gravel during placer gold mining operations in the Ural Mountains in the late eighteenth century. But the first hard-rock deposit of diamonds in Russia – the type embedded in large rock formations and retrieved by mining – was not discovered until June 13, 1955. On that day Yuri Khabardin, a geologist working in Yakutia, Siberia, accidentally stumbled upon a geological outcrop known as a "volcanic pipe" that contained diamond crystals frozen in a bluish igneous rock called kimberlite. Khabardin attributed his spectacular discovery to the fact that he had happened upon "a fox who had thrown the bluish kimberlite earth out of its lair."

Khabardin's diamond deposit, known today as the Mir Pipe, has since produced billions of dollars worth of

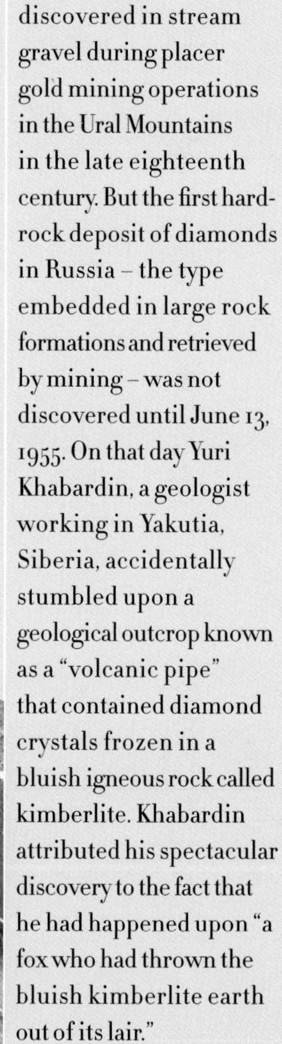

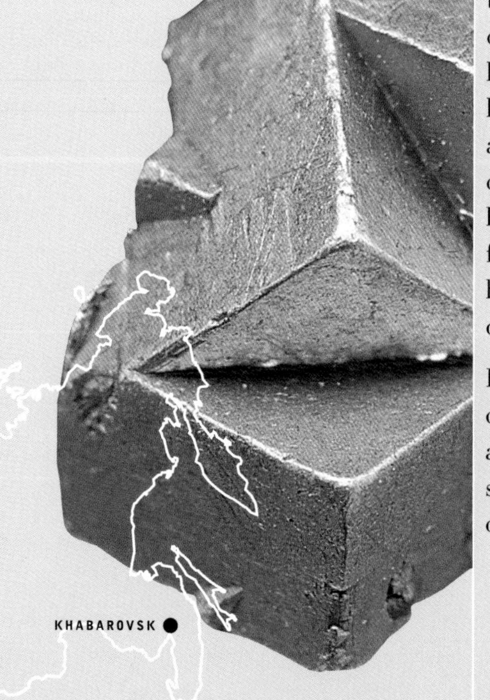

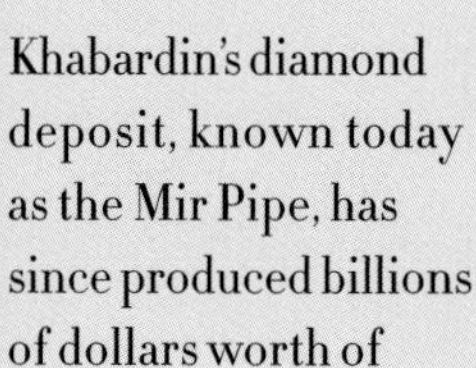

transparent diamond crystals of the finest quality. In December 1981 the deposit produced a single diamond crystal weighing 342.5 carats, which was later dubbed "The 26th Congress of the Communist Party of the USSR Diamond." Many other enormous crystals have been found at the Mir Pipe, including the 320.6 carat "Alexander Pushkin Diamond" recovered in 1989. By order of the Russian government, diamond crystals weighing more than 20 carats are preserved in their native state and added to the treasury. Diamond crystals weighing less than that are fashioned into gems and sold on the world market.

YAKUTIA

RUBY

Ruby and sapphire are varieties of the mineral corundum (aluminum oxide). For a corundum crystal to be considered a true ruby it must be transparent or at least highly translucent and possess a deep red "pigeon blood" body color caused by the presence of chromium. Though always rare, fine-quality rubies became more readily available beginning in the eighteenth century and were especially favored by Russian jewelers working in St. Petersburg. An example of their work is the gold-mounted snuff box with ruby flowers produced around 1850 [*plate 70*].

Red corundum crystals up to 20 centimeters in length have been found in the Rai-Iz massif in the polar Urals. Typically opaque and too highly fractured to be faceted as gems, some have been used for carvings or polished as natural, uncut "cabochons." Although stones were also mined at Jagdalek in present-day Afghanistan, the vast majority of the rubies used in both Western and Eastern European cultures over the centuries originated in the Mogok gem mines of Burma (now Myanmar).

SAPPHIRE

Sapphire is the name given to gemmy crystals of corundum (aluminum oxide) that are commonly colored blue by trace amounts of titanium and iron, though sapphires also occur in other colors, including pink, orange, purple, and yellow. During the seventeenth century, pendant-mounted blue sapphires representing the heavens were often suspended from the *panagias*, or pectoral insignias, worn by Russian Orthodox clergy [*plates 44, 31*]. A more recent, secular example of the use of sapphires is in a pendant [*plate 83*] made in Moscow at the beginning of the twentieth century.

While stream-worn chunks of bluish-gray corundum have been found in the river gravel of the Ural Mountains and in the Kola Peninsula, few have proven suitable for faceting into gemstones. Deposits of industrial-grade corundum are also known in Siberia. In 1745, a single grayish-blue sapphire was presented to the tsar; the stone reportedly had been found by a peasant girl at Kornilova in the Ural Mountains. Reports of such isolated discoveries, however, are the exception. Most of the beautiful sapphires that have adorned Russia's jeweled objects over the centuries originated in the gem mines of Ratnapura in Ceylon (now Sri Lanka).

KOLA PENINSULA

POLAR URAL MOUNTAINS

EMERALD

Emeralds are crystals of the mineral beryl (beryllium aluminum silicate) that exhibit a bright grass-green color caused by the presence of minor amounts of chromium or vanadium. Through the fifteenth century, the only emeralds available in Europe were Egyptian stones of inferior quality. This changed in the sixteenth century when spectacular emeralds began to appear on the market from a mysterious "lost" mine in India. Actually, all of the new emeralds were originally from the mines around Muzo in Colombia. The gems had found their way to Russia through brokers in Europe and India, who, in turn, had acquired them from Spanish merchants returning from the New World. The icon cover known as "Our Lady of Odighitria" [*plate 18*] contains emeralds attributed to the Colombian mines.

Though it is possible that the "Scythian emeralds" mentioned by Pliny the Elder in 79 AD were from the Ural Mountains, it is commonly accepted that the first discovery of gem-quality emeralds in Russia was made on January 23, 1831, near Sverdlovsk, by a peasant named Maxim Kozhevnikov. While inspecting the condition of his alcohol-distilling works after a particularly violent storm, Kozhevnikov noticed unusual green crystals nestled among the roots of an overturned tree. Although mistakenly identifying the stones as aquamarines of inferior color, Kozhevnikov brought them to the attention of the director of the Imperial Lapidary Workshops at Ekaterinburg, Yakov Kokovin, who recognized their value and immediately began emerald mining operations. Kozhevnikov's discovery, for which he received an award of 200 rubles, led to the development of one of the world's most prolific emerald mines, which still operates today as the Sretenskiy Mine at Izumrudnye Kopi.

Kokovin's four-year tenure as director of operations at the mine, however, began gloriously and ended ignominiously. His first two years of successful mining enabled him to present

Tsarina Alexandra Feodorovna with a faceted, pear-shaped gem weighing 101 carats, for which he was awarded the prestigious Order of St. Vladimir. Shortly thereafter, a second magnificent emerald, a rough crystal weighing 2.226 kilograms, was found at the mine and acquired by Prince Kochubei for his personal collection. Unfortunately for Kokovin, a third, equally famous crystal weighing almost 500 grams and "surpassing in merit the emerald once mounted in the crown of Julius Caesar" mysteriously vanished. Though he proclaimed his innocence, the unexplained disappearance cost Kokovin his job and his life.

AQUAMARINE

Aquamarines are transparent crystals of beryl (beryllium aluminum silicate) ranging in color from pale sea-green to deep sky-blue because of trace amounts of ferric iron in their crystal structure. In Russia, aquamarine has been popular for centuries for the decoration of icon frames and other religious objects. The 1589 "Reliquary Icon of Tsarina Godunova" [*plate 20*] is surmounted by a flawless pale-blue aquamarine of unknown origin.

Aquamarine crystals were originally discovered in Russia at Adun-Chilon in 1723 by a schoolteacher from Nerchinsk. Although he incorrectly identified the crystals as "greenish tourmaline," officials at the State Mining Academy awarded him five rubles for his discovery. The term aquamarine was first used by the Swedish mineralogist Jakob Wallerius to distinguish the transparent crystals of gem-quality beryl found at Sherlovaya Gora in Russia's Transbaikal region. Aquamarine was also found in the Aduy-Mursinka region at the end of the seventeenth century and in East Transbaikalia in the mid-eighteenth century. In 1828, a 2.507-kilogram gem-quality crystal of aquamarine was found at Startseva Iama, near Alabaschka, in the middle Urals. It was valued at 150,000 rubles at the time and transferred to St. Petersburg for "safekeeping" after the Imperial Court was informed that such a stone had never been found before. In recent years, medium-blue gemstones have been produced for the retail market by heating the yellowish-green crystals of so-called heliodor beryl found at the mines near Volyn, Volodarsk-Volynskiy, in Ukraine.

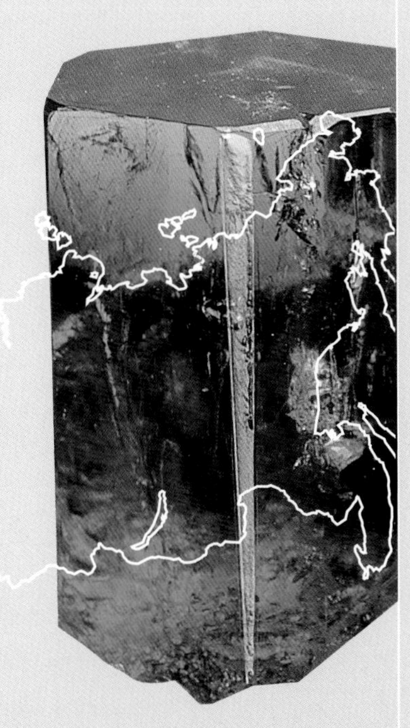

SVERDLOSK

STARTSEVA IAMA

NERCHINSK

TRANSBAIKALIA

ALEXANDRITE

Alexandrite is perhaps the most "Russian" of all gems: It was first discovered in Russia, was not found elsewhere for over a century, and was named for a Russian tsar. A transparent variety of the mineral chrysoberyl (beryllium aluminum oxide) that changes color in different types of light, alexandrite appears reddish-purple in incandescent light but changes to bluish-green in daylight. Although the mineral chrysoberyl is fairly common, transparent, flawless crystals that exhibit the so-called "alexandrite effect" are among the rarest of all gems.

Alexandrite was first discovered in the Ural Mountains at the Izumrudnye Kopi, near Ekaterinburg, in the same locality that produced Russia's finest emeralds. The Finnish mineralogist Nils Nordenskiold named the new gemstone "alexandrite" to commemorate the fact that the date of discovery, April 17, 1834, was the sixteenth birthday of Alexander, the eldest son of Tsar Nicholas I, who would assume the throne in 1855. Crystals of the alexandrite variety of the mineral chrysoberyl are usually quite small. Although a few stones weighing as much as 30 carats are known, alexandrites weighing even five carats are extremely rare. Between 1880 and 1916, Carl Fabergé assembled one of the finest collections of Russian gemstones ever seen, yet it contained only two faceted alexandrites, each weighing less than three carats.

GARNET

ALMANDINE

Found in abundance throughout Eastern Europe, garnets have been used for centuries to decorate jeweled objects of every description. Wine-red in color, almandine garnets (magnesium aluminum silicate) were popular adornments on objects created for Russian Orthodox churches. An early Byzantine example of the liturgical use of almandine garnets is found in the eleventh-century icon "Christ Giving Blessings" [*plate 4*], in which almandine garnets are mounted in the gold frame that surrounds the icon. Garnets were also used as accent stones in the twelfth- to thirteenth-century gold *barmy*, or collar [*plate 11*], and the gold *kolt* medallion [*plate 12*] made during the same period.

GARNET

ANDRADITE-DEMANTOID

The most famous of Russian garnets are the exceedingly rare, bright green demantoid garnets (calcium iron silicate) found at Bobrovskoye and Poldenovskoye in the Ural Mountains. Demantoid garnets take their name from the German "diamant," in reference to the spectacular, diamond-like brilliance of the polished gemstones. The intense, fiery green color of demantoid garnets is caused by trace amounts of iron and chromium in the crystal structure.

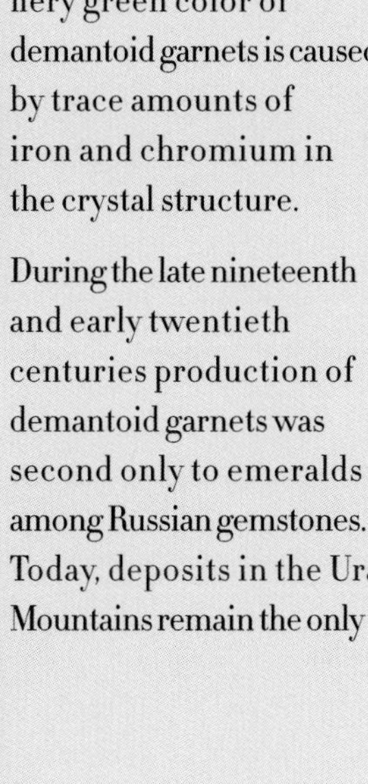

During the late nineteenth and early twentieth centuries production of demantoid garnets was second only to emeralds among Russian gemstones. Today, deposits in the Ural Mountains remain the only Russian source for high-quality stones, although smaller crystals, unsuitable for cutting, have been found in Kamchatka and on the Chukotka Peninsula. In the last few years, renewed mining in the Ural Mountains has provided the world market with magnificent gems weighing up to 30 carats. Large crystals of demantoid garnet also have been found in Italy; their brownish-green tint, however, makes them appear less lively than their Russian counterparts.

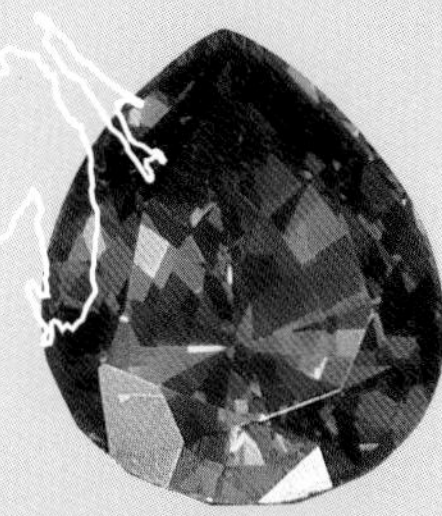

"ROCK CRYSTAL" QUARTZ

Known as rock crystal, transparent, colorless crystals of quartz (silica dioxide) have been used since antiquity as ornamental stones in jewelry and other objects. The central stone, for example, in the twelfth- to the thirteenth- century gold *kolt* medallion [*plate 12*] is a large piece of polished quartz. During the seventeenth century, Russian craftsmen used rock crystal quartz to create elegant vases and drinking vessels, an outstanding example of which is the gold-mounted quartz *charka*, or wine-tasting cup, made in the Moscow Kremlin Workshops in 1636 [*plate 26*]. At the beginning of the twentieth century, Carl Fabergé used a transparent piece of quartz as the central element in one of his fabled Easter eggs for Tsar Nicholas II [*plate 92*].

Since the late 1920s many tons of flawless quartz crystals have been found in the Dodo and Puiva regions of Russia's polar Ural Mountains. Many of the crystals were used as raw materials for Russia's radio and quartz glass industries and in the manufacturing of fiber-optic communication systems. In addition to industrial applications, flawless crystals of rock crystal quartz from the Urals continue to be prized by collectors and connoisseurs, as well as by lapidary artisans around the world.

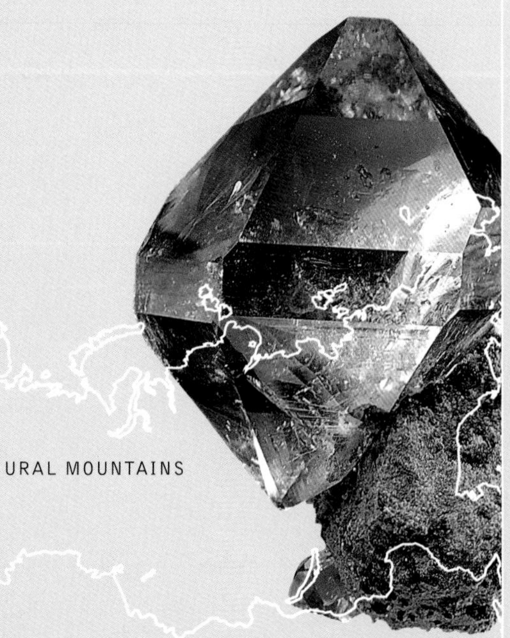

AMETHYST

Amethyst is the purple variety of quartz (silica dioxide), which owes its color to trace amounts of ferric iron in its crystal structure. The name amethyst is derived from two Greek words meaning "no alcohol," a reference to the ancient belief that, by virtue of the wine-like hue of the gem, anyone who wore an amethyst or placed it in his goblet could drink copious amounts of wine without becoming intoxicated. In Russia, amethyst deposits are known in Siberia, Irkutsk, Khabarovsk, the Ural Mountains, and the Kola Peninsula.

Amethyst has been an immensely popular gemstone throughout Europe and in Russia for most of the last two centuries, and has been used to adorn a wide variety of precious objects. The amethyst mines at Talyan, near Mursinka, were opened in 1768 and produced an abundant supply of high-quality stones throughout the late eighteenth and early nineteenth centuries. One late-nineteenth century author, D. N. Mamin Sibiryak, reporting from the amethyst mines near Murzinka wrote:

The most salable stone among the Murzinka gems is amethyst, which has penetrated far into Europe and has a firm hold on the market. . . . This is connected with the appointment of several new Orthodox bishops, because, mounted in icons, miters, and finger crosses the gem is most effective. In this regard, amethyst may be termed "the stone of the Orthodox church," just as green tourmaline is the favorite stone of the Catholic clergy.

A seventeenth-century example of the Russian affinity for amethyst is the high-relief, chased gold sarcophagus cover of Tsarevitch Dmitry Ivanovich [*plate 28*], the youngest son of Ivan the Terrible, who was murdered as a young boy in 1591, supposedly at the order of Boris Godunov. The life-sized figure of Dmitry on this cover is clothed in golden robes accented with faceted amethysts. A more recent example of the Russian love of amethyst is found in the diamond crown [*plate 114*] created in 1990 to celebrate the 500th anniversary of the Kremlin's Annunciation Cathedral. The central elements of the design are four large, faceted amethysts mined in the Ural Mountains. Amethyst from the tsar's mines in the Urals, known as "imperial amethyst," is widely regarded as the finest in the world because of red and blue highlights in the deep purple body color of the stones.

CITRINE

Citrine, the yellowish variety of quartz (silica dioxide), has been prized for centuries as a gemstone in its own right, though it also has been used, sometimes deceptively, as a less expensive substitute for orange-yellow imperial topaz. Although crystals of citrine quartz have been found in Russia's polar Ural Mountains, most citrine gemstones are produced by heating crystals of amethyst in a furnace to change their color from purple to yellowish-orange.

For centuries citrine has been given names suggesting it is a rare variety of topaz, imported from a mysterious and exotic source. In the seventeenth and eighteenth centuries, citrine was commonly known in Europe as Indian topaz, since the yellow stones had once been known to occur in India. Evidence of this belief is found in an alphabetical gemstone cipher mounted across the lid of a gold and diamond snuffbox of the early nineteenth century [*plate 66*], on which faceted citrines, or Indian topaz, are used to represent the letter "I."

SMOKY QUARTZ

Smoky quartz, often erroneously referred to as smoky topaz, was first extracted in 1720 from quartz veins in the Ural Mountains that outcropped along the Neiva River. For more than one hundred years stones from this locality were incorrectly identified as topaz. In the twentieth century tons of smoky quartz have been produced from mines near Puiva, in Russia's polar Ural Mountains.

Rarely used in Russian jewelry of the eighteenth and nineteenth centuries, Russian smoky quartz enjoyed new prestige late in the nineteenth century when it was used in the creations of imperial jeweler Carl Fabergé. Artisans working for the Fabergé firm used smoky quartz for everything from simple animal carvings to elegant gold-mounted vases. Fabergé's clever use of smoky quartz in his creation of the Imperial Easter Egg for 1900 [*plate 91*] is a testimony to his attention to detail. Hidden inside the egg is a working model of the Trans-Siberian Railway train executed in gold. The window panes in the smoking sections of the passenger cars are made from thinly sliced, highly polished pieces of smoky quartz, while those in the non-smoking sections are made of colorless rock crystal quartz.

URAL MOUNTAINS

TOPAZ

Deposits of gem-quality crystals of topaz have been mined throughout Russia and the surrounding territories for at least three centuries. The most rare and valuable are the intensely pinkish-red crystals found along the Sanarka and Kamenka Rivers in the southern Urals. Though the color is superb, the crystals tend to be somewhat small for faceting as gemstones. The large, pale pink gems found in Russian jewelry of the nineteenth century were often imported from Brazil. Russian jewelers referred to these stones as "burnt topaz" to indicate that the stones had been heated in a furnace to induce the pink color.

Flawless crystals of colorless topaz were first found in the Transbaikalia region at Adun-Chilon in 1723 and later along the Urulga River in the Borschovochniy Mountains. Brownish crystals weighing several kilograms have also been found in the deposits at Volodarsk-Volynskiy in Ukraine. Russia is perhaps most famous for the perfectly formed, internally flawless blue topaz crystals found at Mokrusha, which weigh as much as 14 kilograms. These magnificent crystals prompted one nineteenth-century gemologist, Nils Nordenskiold, to write in 1850:

Russia may indeed be proud of its topazes, which in terms of clearness of water and size of crystals occupy an exclusive place among the best topazes of the world.

A remarkable golden topaz in this exhibition is used as a signet stone [*plate 81*] with three polished surfaces, each engraved with a different insignia. Among other gemstones, topaz is used in a row of stones lining the lid of the nineteenth-century diamond, gold, and blue enamel snuffbox [*plate 66*].

SOUTHERN URAL MOUNTAINS

TOURMALINE

The name tourmaline refers to a complex group of silicate minerals that includes numerous types of crystals, many of them brownish-black and opaque. However, some species of tourmaline, such as elbaite, occur as transparent, lustrous, prismatic crystals in a wide variety of colors. Most of the transparent elbaite crystals found in Russia are known as "rubellites" because of their red color. In 1810, brothers Fedor and Afanasiy Kuznetsov found loose crystals of gem-quality rubellite in river gravel near the Ural Mountain village of Sarapulka. In 1815 mining began under the direction of Jakob Mor, director of the Ekaterinburg Imperial Lapidary Factory, who reported that he was "extremely happy to have encountered very rich nests of this rare gemstone."

The Russian people have long held rubellite tourmaline in high regard, even before domestic discovery of the gem. Evidence of this esteem is seen in the magnificent diamond crown made for the coronation in 1682 of Tsar Ivan V [*plate 35*]. The central element of this priceless crown is a naturally shaped, uncut rubellite of superb color that weighs at least 100 carats. The stone probably was acquired in Asia and brought to Moscow to be added to the gem collection in the Imperial Treasury.

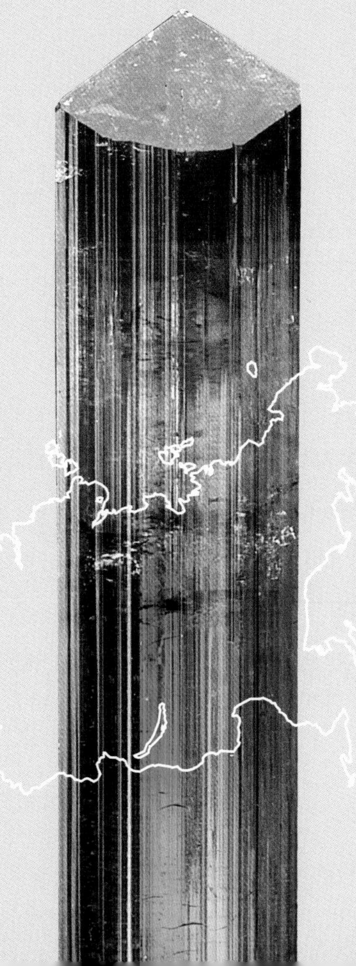

AGATE

The term agate refers to a translucent, multicolored variety of fine-grained quartz whose colors often exhibit a layered or banded structure. Agate has been known from sources in Kazakhstan, Georgia, Azerbaijan, and Armenia since medieval times. Engravers in Eastern Europe and in Mediterranean countries have used naturally banded agates for centuries to produce high-relief cameo carvings. Exploiting the different colors in the alternating layers of naturally banded agates, artisans created images with contrasting colors which heightened the three dimensional effect of the cameo carving.

Portraits of mythical heroes, biblical figures and political leaders were the most common images executed in agate, although more complicated allegorical scenes were also produced. An outstanding sixteenth-century example is the "Angel of the Desert" *panagia* [*plate 20*], or pectoral insignia, carved from a single piece of banded agate found near Venice, Italy. In Russia today, high-quality agates are found in the regions of Timan, Primorsky, and Transbaikalia, as well as in the Ural Mountains.

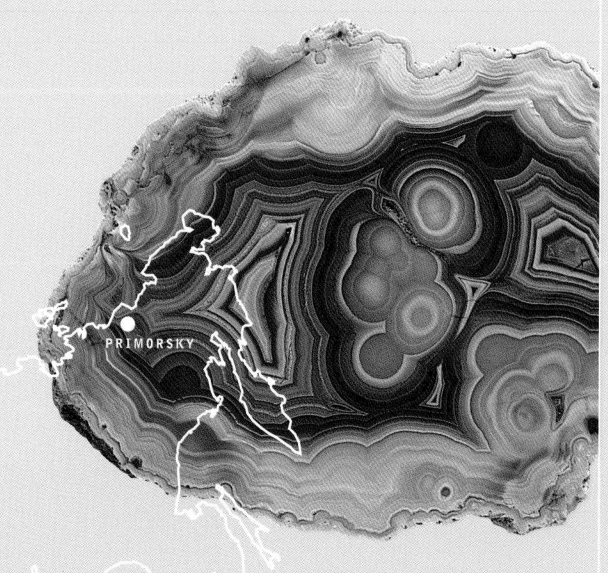

JASPER

Jasper is another variety of multicolored, fine-grained quartz. Although it tends to be opaque and its color zoning is less regular than that of agate, jasper was also used to make cameos and other decorative objects. In the eighteenth and nineteenth centuries, polished slices of blotchy, multicolored jaspers Mountains were mounted in frames because of their resemblance to landscape paintings.

In 1829 a block of gray-green jasper weighing more than twenty tons was discovered at the Revnevskoya deposit, near the village of Zmeinogorsk in the Altai Mountains. More than two years were spent excavating the block and more than twelve devoted to carving it into a vase five meters high and three meters wide. The piece was transported to St. Petersburg using a team of 120 horses. From start to finish the project required the efforts of more than 700 workers. An earlier and much smaller example of the use of jasper as a decorative stone is the eleventh-century cameo icon "Christ Giving Blessings" [*plate 4*], which features an image of Christ carved in bas-relief. This piece of jasper, of a type commonly called "bloodstone," was no doubt chosen because of the presence of crimson blotches representing Christ's blood.

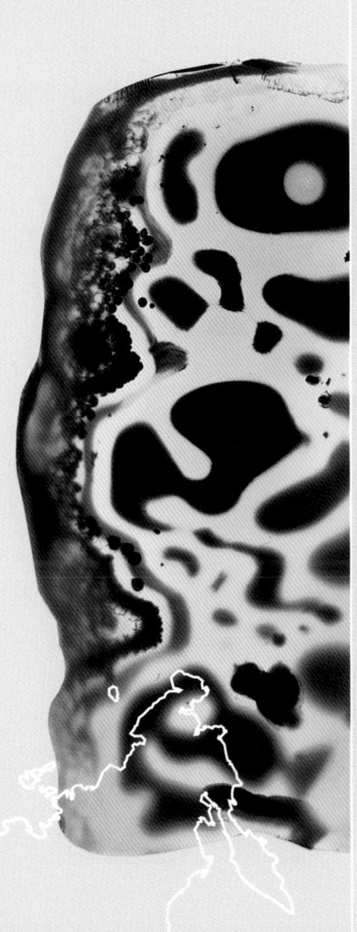

CARNELIAN

Another red form of fine-grained quartz related to both agate and jasper is carnelian, whose uniform color and warm translucency made it a popular ornamental stone among the Russian clergy. An exceptional example of the use of carnelian in liturgical jewelry is found in the seventeenth-century *panagia* [*plate 44*] created for bishops of the Russian Orthodox church in Moscow. The polished surface of the hexagonal stone features an intaglio carving of the image of Christ. In the twentieth century Fabergé used carnelian for producing hardstone animal carvings such as the dolphin [*plate 86*] sold by the St. Petersburg branch of his firm around the turn of the century.

MALACHITE

The mineral malachite (copper carbonate hydroxide) derives its name from the Greek term *malache*, meaning "bright green plant," an allusion to its color and fibrous structure. Although relatively soft, malachite is able to take a fairly high polish, making it a popular ornamental and decorative stone. In 1702 an enormous deposit of velvety green malachite was discovered in an area of the Ural Mountains that came to be known as the Gumeshevskiy copper mining region. Though archeological evidence suggests the region was mined for copper as early as 2000 B.C., it was not until malachite became popular for decorating palace interiors during the eighteenth and nineteenth centuries that the mines began to flourish. In 1789 a 1,500-kilogram block of malachite was extracted from the Gumeshevskiy mine. On receiving the news of its discovery, Catherine the Great ordered that it be preserved as a national treasure and it was subsequently moved to St. Petersburg, where it remains today.

URAL MOUNTAINS

LAPIS LAZULI

Lapis lazuli is the popular name given to dark blue pieces of lazurite (sodium calcium aluminum silicate). It often contains gold-colored specks of pyrite (iron sulfide) and white veins of marble (calcium carbonate). Lapis lazuli has been a extremely popular stone in Eastern European cultures since antiquity. In this century, its increased availability has made it more widely accepted in the Western cultures, as well. Merging Eastern and Western tastes, the Russian jeweler Carl Fabergé used lapis lazuli as a major element in the "Standart" Imperial Easter Egg [*plate 92*] made in 1909 for Tsar Nicholas II. The colorless quartz egg is flanked on either end by lapis eagles and rests on the tails of lapis dolphins.

Since antiquity, the most productive source of the finest quality lapis has been the Sary-Sang deposit in present-day Badakshan, Afghanistan. For centuries, stones from this area have been prized for their rich, uniform color and were usually reserved for use in important pieces of jewelry. Stones from other places, typically possessing a lighter color because of impurities, were used in mosaics and other decorative objects, or crushed into a fine powder for use in ultramarine dyes.

In 1787, 320 kilograms of lapis lazuli were extracted from a quarry in southern Irkutsk near Sludyanka and shipped to St. Petersburg, where the stone was later used to decorate the interior of the imperial Tsarsko-Selo palace. Lapis lazuli has also been recovered from the Malaya Bystraya River in Russia's southwestern Pribaikalia region and in the Pamir Mountains of Tadzhikistan in chunks weighing up to ten kilograms.

RHODONITE

One of the most popular decorative stones in Russia, rhodonite (calcium manganese silicate) was first discovered in 1795 by a peasant from the village of Malaya Sedelnikovaya, near Ekaterinburg. At the time, all such discoveries of precious stone deposits immediately became the property of the tsar. Mining operations in this region thus were carried out in the name of Catherine the Great and her successor, Paul I, under the supervision of the director of the Imperial Ekaterinburg Lapidary Workshops.

In 1869 miners extracted a single block of rhodonite weighing more than 45 tons from the Bolshaya Orletzovaya mine. This imposing piece accounted for nearly one-third of the rhodonite mined in the region that year. The block was preserved for twenty years before being carved into a sarcophagus for Maria Alexandrovna, the wife of Tsar Alexander II. It was installed in the Petropavlovskiy Cathedral in St. Petersburg in 1906.

On a much smaller scale, the Fabergé firm used rhodonite for such useful objects as the cigarette case and matching cufflinks [*plate 73*] made for Tsar Alexander III during the last quarter of the nineteenth century. The finest rhodonite is devoid of impurities and ranges in color from pale grayish pink to a rich crimson red, and was admirably suited to the creation of Fabergé's signature luxury items. In 1917 the rhodonite mines were virtually abandoned and remained dormant until the mid-1930s, when the quarries were reopened to provide slabs for decorating the interior of the Mayakovskaya railway station then under construction in Moscow. For the next fifty years the mine was worked only sporadically. In 1989 the quarries were filled in and leveled to reclaim the land for agricultural use.

EKATERINBURG

IRKUTSK

NEPHRITE

The term jade refers to two distinctly different minerals, jadeite (sodium aluminum silicate) and nephrite (calcium magnesium iron silicate). Nephrite occurs most commonly as a dark, spinach-green stone with opaque black flecks scattered throughout. Nephrite also occurs in colors ranging from pale green to bone white, though these are much rarer than the dark green variety. In the summer of 1851, boulders of dark green nephrite weighing more than 2,000 kilograms were found along the Onot River in eastern Siberia. Russian nephrite came to be favored by Carl Fabergé, who used it to produce carvings, letter openers, monogrammed jewelry boxes, and other luxury items for his wealthy clients. The recent "Stele with Portrait of Fabergé" [*plate 116*] is made of Russian nephrite.

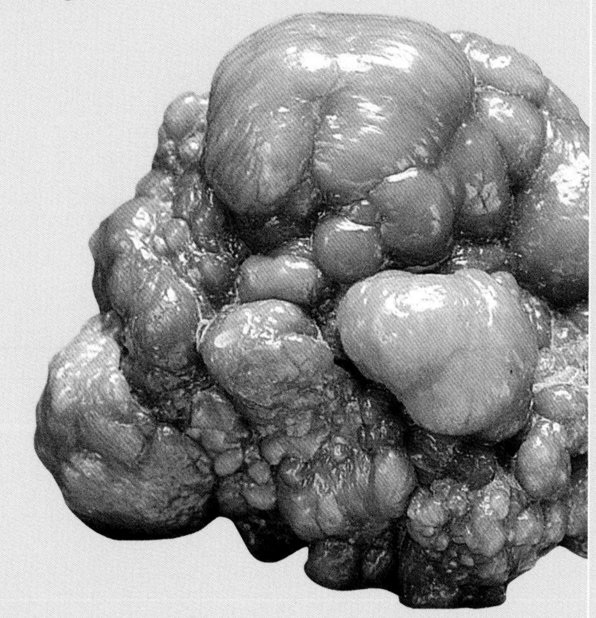

JADEITE

High-quality jadeite (sodium aluminum silicate) is far rarer than nephrite and occurs in more colors, including red, orange, yellow, lavender, green, and black. The rarest and most highly prized variety of jadeite is "imperial" jade, a highly translucent green stone with a color rivaling that of the finest emeralds. In Russia, jadeite has been recovered from mines in the polar Urals and in western Sayan. Russian objects made of jadeite are uncommon because until the 20th century, its primary source was Burma (now Myanmar).

POLAR URAL MOUNTAINS

SAYAN

Letter on Russia of the 16th Century, Journal of the Ministry of Public Education, 1842, no. 9, pp. 148–49; quoted in *Precious Stones in Russian Jewelry Art,* M. V. Martinova, Moscow, 1973.

"Readings in the Society of History and Antiquities of Russia," 1884, vol. 1, p. 225; quoted in *Martinova*, Moscow, 1973.

I. Zabelin, *Home Life of the Russian Tsarinas in the 16th and 17th century,* Moscow, 1872, p. 592; quoted in *Martinova*, Moscow, 1973.

"Proceedings of the Russian Imperial History Society," 1884, vol. 41, p. 40; quoted in *Martinova*, Moscow, 1973.

Precious Stones in Russian Jewelry Art, *Martinova*, Moscow, 1973.

Michael Leybov, *World of Stones*, Moscow, 1990–98.

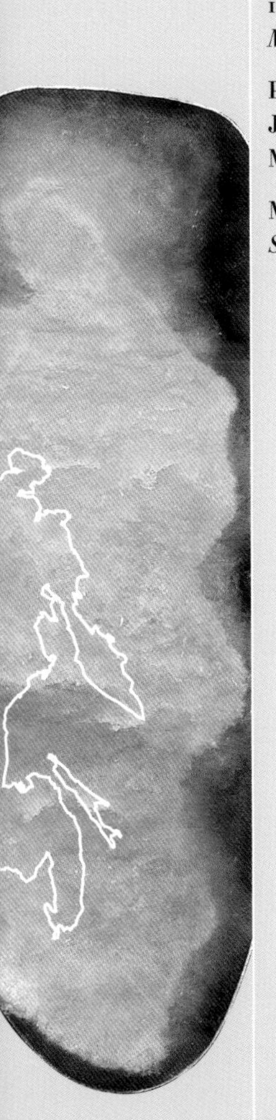

THE GEOGRAPHY OF RUSSIA

GEOGRAPHY

Stretching across parts of two continents and spanning eleven time zones – almost half the circumference of the earth – Russia is a land of vast spaces and dramatic extremes of climate and topography. From the frigid islands of Franz Joseph Land in the Arctic Ocean to the subtropical shores of the Black and Caspian Seas in the south, from the fertile farmlands of the west to the barren realms of far eastern Siberia, Russia's geography sets a stage as rich and varied as the people who have called this land home.

Occupying nearly one-sixth of the Earth's land mass, an area more than twice as large as either the United States or China, the countries of the Confederation of Independent States are for the most part sparsely settled and offer inhospitable climates. Much of the territory lies at latitudes north of the U.S.-Canadian border, and is covered by treeless tundra or the thin evergreen forests known as the *taiga*. In the south, seas of grass – the legendary Russian steppes – gradually give way to deserts and the encircling mountains. Russia's population, agriculture, and industry historically have been concentrated in the so-called Fertile Triangle, a broad swath of rich, well-watered land that sweeps in a triangular arc from the plains of Central Asia to the breadbaskets of western, European Russia.

Russia's many-sided geography has shaped its history since prehistoric times. Straddling Europe and Asia, it has been an eternal crossroads for populations on the move. The same Fertile Triangle that nurtured and sustained the Slavic peoples of Old Russia also offered an open road for invasions by nomadic tribes from distant East Asia, notorious among them the Huns and the Mongols. And in a boundless land by turns frozen in winter's grip, awash in the spring thaw, or covered by a sea of shoulder-high grass, Russia's mighty rivers have provided the highways down which people and goods have moved over the centuries.

SALINE LAKES

Russia's saline lakes are among the largest enclosed bodies of water on Earth. Greatest in size and in the diversity of nations and ethnic groups that hug its shores is the Caspian Sea, the world's largest lake as measured in surface area. Like Russia's other saline lakes, it is fed by freshwater rivers but does not overflow to the sea. Water escapes only through evaporation, leaving behind salts that accumulate and make the lake highly saline.

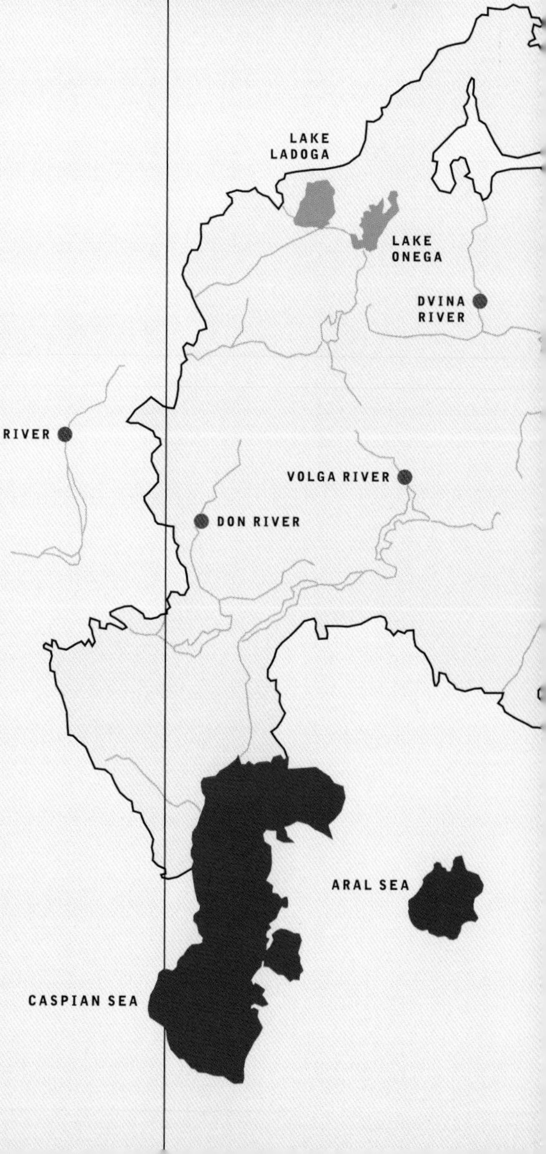

LAKE LADOGA
LAKE ONEGA
DVINA RIVER
DNIEPER RIVER
VOLGA RIVER
DON RIVER
ARAL SEA
CASPIAN SEA

RIVERS

Russia's rivers (the Don, Dnieper, Dvina, and Volga in the west; the Ob, Irtysh, and Yenisei in the central plain; the Lena and Amur in the east) are among the largest and longest in Europe and Asia. Forming great natural roads in ancient days – Scandinavian Vikings coursed down the Dnieper in the ninth century to found their capital at Kiev – Russia's rivers in modern times have been harnessed for hydroelectric power projects and the transport of industrial goods.

FRESHWATER LAKES

Foremost among the many freshwater lakes fed by Russia's rivers and filling scars gouged by ancient glaciers is Lake Baikal. This pristine body of water deep within the barren realms of south-central Siberia is the world's deepest freshwater lake and the largest in terms of volume, holding one-fifth of the fresh water on Earth.

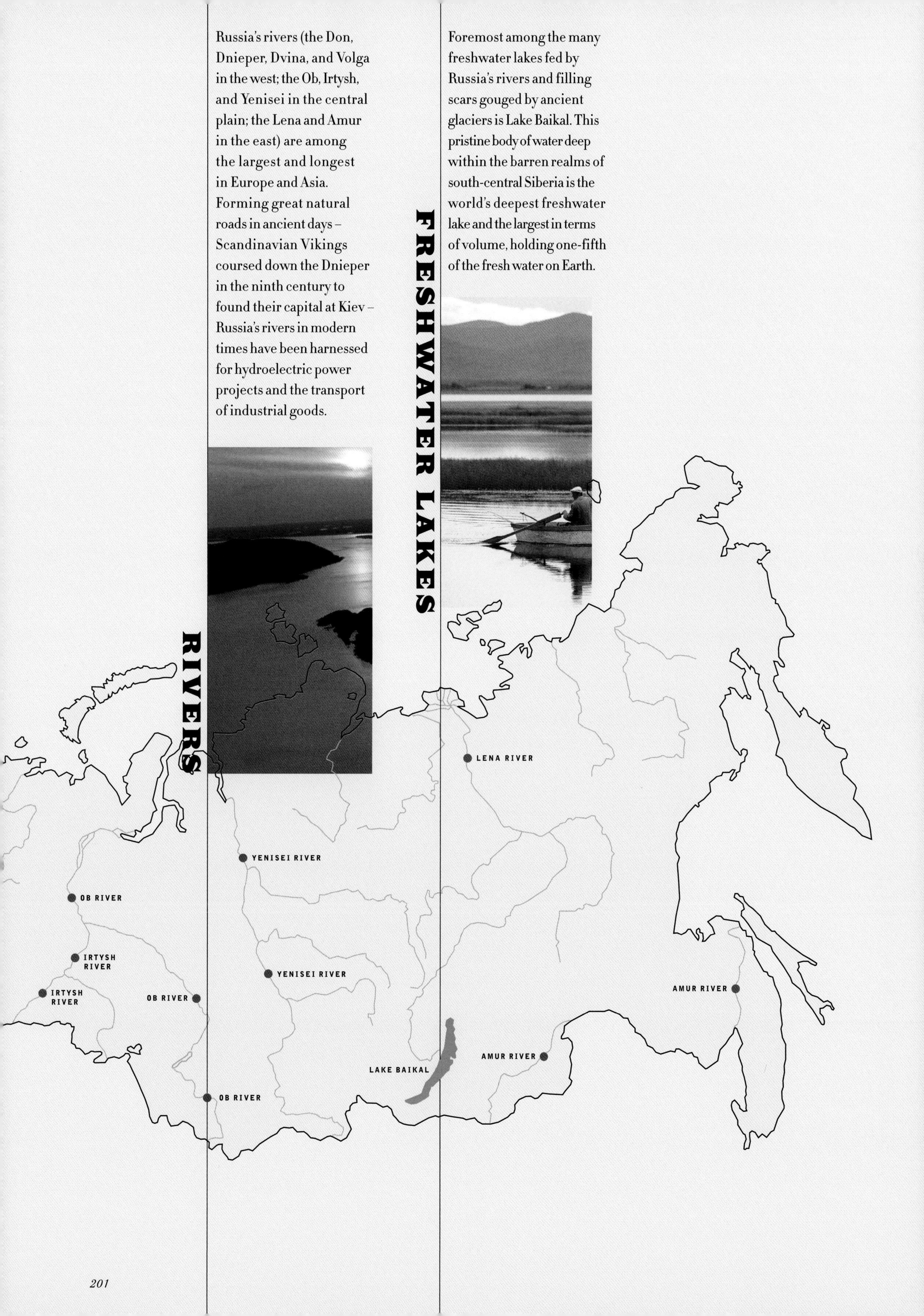

SOUTHERN STEPPES

In the rich black soil of the Russian steppes, summer grasses grew high enough to reach the knees of a mounted horseman. Sweeping across southern Russia from Central Asia to the threshold of Western Europe, the steppes have been Russia's breadbasket since ancient times, providing ideal conditions for the cultivation of grains. This same luxurious golden sea of grass, however, also beckoned the nomadic tribes who invaded Russia in successive waves over the centuries: Scythians, Sarmatians, Huns, Avars, Pechenegs, and finally the fierce Mongols.

NORTHERN FOREST

The forests blanketing Russia's northern half offered poor provender to the Slavic peoples who migrated here from southeastern Europe before the ninth century A.D. Although nutrient-poor clay soils could not sustain the same bountiful agriculture that graced the southern plains, the forests provided many marketable products, such as furs, honey, and beeswax, which encouraged trade with the south. Dense forests also protected northern dwellers from the ravages of nomadic invasion. Here the Russian craftsman nurtured his native genius for woodworking: the famous onion-domed wooden churches of Old Russia were often shaped with no tool but an axe.

DESERTS

Fabled cities on the trade routes through Central Asia – Tashkent, Astrakhan, Samarkand – pepper the harsh desert zone on Russia's extreme southern borders. Here, where Slavic Russia gradually yields to a host of diverse ethnic groups and cultures, dry cold winters, hot summers, and barren sandy soils discourage all but the hardiest plants and animals. Horsemen and herders have long held sway here, where even today ancient lifeways and customs linger.

TAIGA

Blanketing the Russian landscape south of the treeless tundra, a vast evergreen forest, the *taiga*, stretches nearly 4,000 miles across the country in a band 2,000 miles wide at some points. Although it holds nearly a third of the earth's softwood timber, the taiga's standing bogs and nutrient-poor soil, called *podzol*, discourage settlement or cultivation. Squirrels, hare, fox, and ermine thrive here, but few humans call the taiga home.

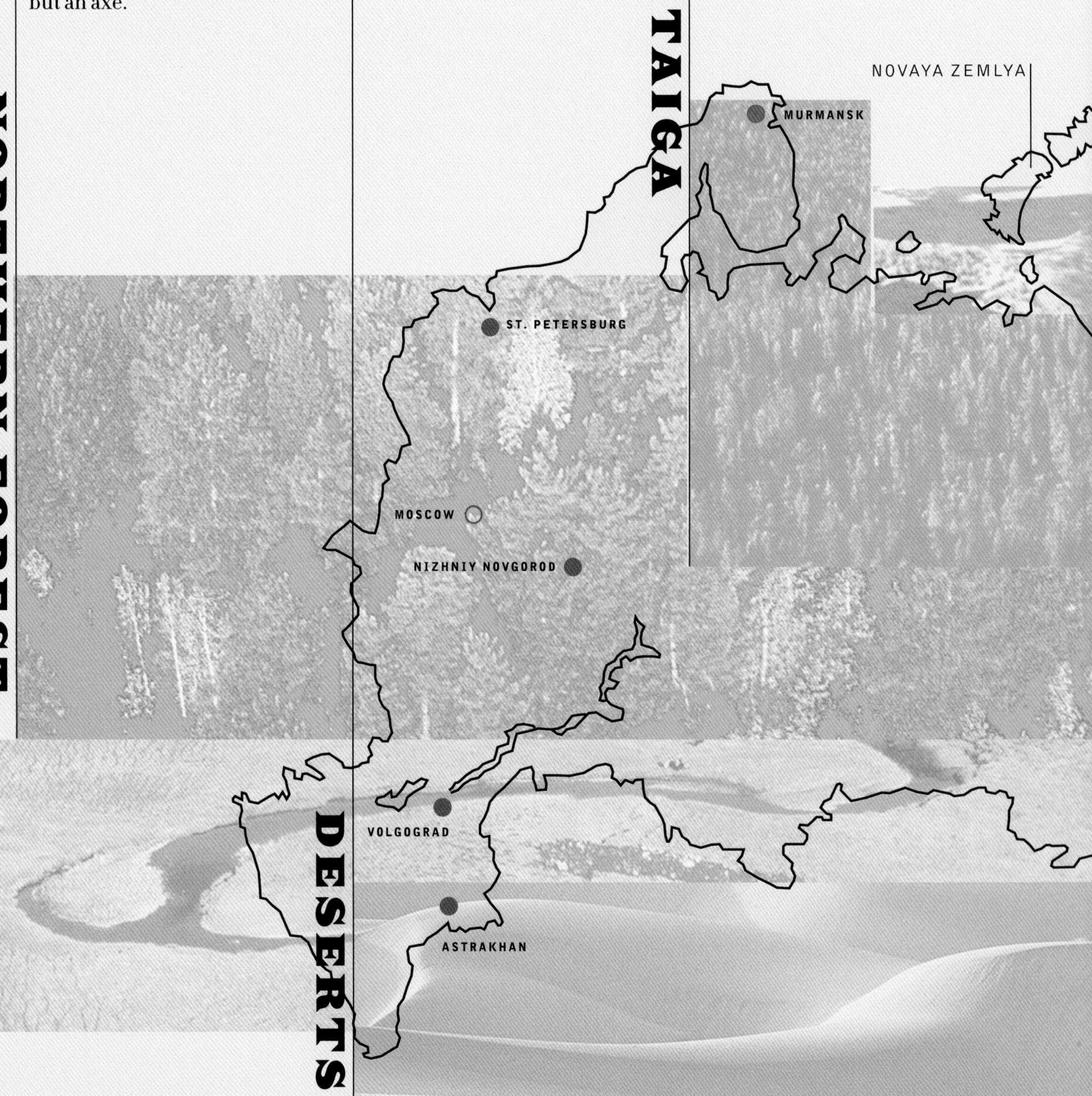

ARCTIC TUNDRA

Far in the north are the polar desert zones of Russia's Arctic islands, among them Franz Joseph Land and Novaya Zemlya. Arctic fox, hare, lemmings, and polar bear vie for survival in a bleak landscape where the permanently frozen subsoil, the permafrost, supports little beyond mosses, lichens, and a few flowering plants.

MOUNTAINS

Girding Russia on its southern borders, a series of mountain ranges block the warm maritime tropical air masses that would otherwise penetrate Russia from the south, resulting in harsh, cold winters for the plains and plateaus of the interior. The Caucasus, the Tien Shan, and the Altai ranges provide mountain refuges for cultures far older than the Russian – among them the Georgians, Armenians, Azeri, and Kazakh – where herds graze on mountain slopes as they have for thousands of years.

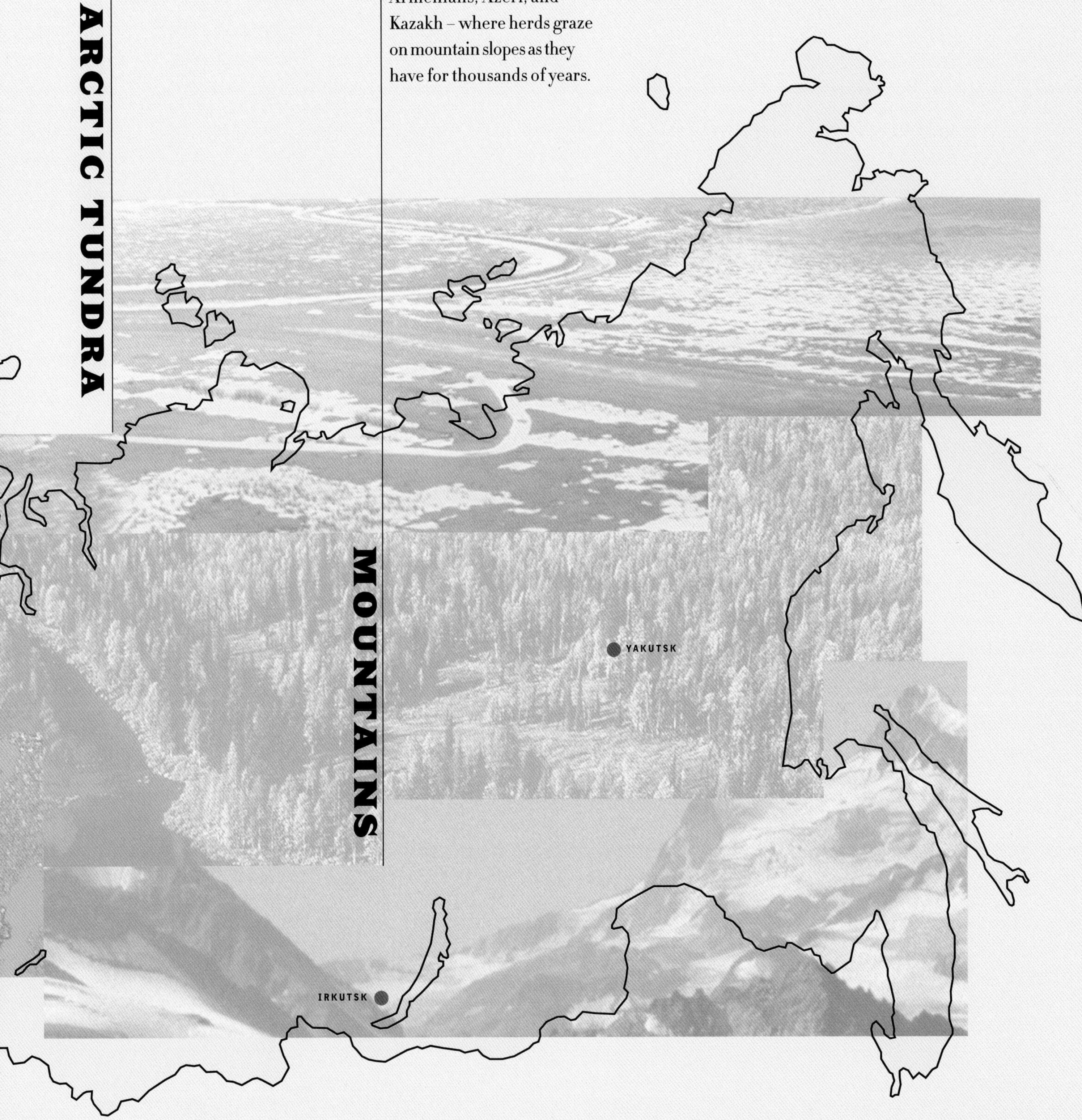

PHOTO CREDITS

All photos by Thomas Dubrock, except as follows:

pages 16–27: figures 1, 2, 4, 6, 7, 14, 16, 17, 18, 20, 21, 28, 36, 43 © Sovfoto/Eastfoto; figures 8, 9, 10, 11, 13, 23, 24, 25, 26, 27, 30, 31, 33, 34, 35, 37, 38, 40, 41, 42, 44, 45, 46, 47, 48, 49 © Corbis.

pages 182–99: Gold nuggets, ruby (courtesy of Michael Scott), sapphire, garnet (andradite-demantoid), citrine, jasper, nephrite, jadeite, topaz (courtesy of Pala Properties) © Harold and Erica van Pelt; native silver, diamond, emerald, alexandrine, garnet (almandine), rock crystal quartz, smoky quartz, tourmaline, agate, carnelian, malachite, lapis lazuli, rhodonite © Jeff Scovil; native gold © Wendell Wilson.

pages 200–203: saline lakes, southern steppes, northern forest, deserts, taiga, arctic tundra © Sovfoto/Eastfoto; rivers, freshwater lakes, mountains © Corbis.

Kremlin Gold: 1000 Years of Russian Gems and Jewels
Joel A. Bartsch and the curators of The State Museums of the Moscow Kremlin.

This book is published in conjunction with the exhibition *Kremlin Gold: 1000 Years of Russian Gems and Jewels*, held at The Houston Museum of Natural Science, Houston, from April 15, 2000 through September 4, 2000 and at The Field Museum, Chicago, from October 21, 2000 through March 30, 2001. The exhibition is jointly organized by the two museums in collaboration with The State Historical-Cultural Museum Preserve, Moscow-Kremlin.

Cover detail of plate 15, pp. 58–59

Published by The Houston Museum of Natural Science and The Field Museum

Distributed in 2000 by Harry N. Abrams, Incorporated, New York
ISBN 0-8109-6695-6

HARRY N. ABRAMS, INC.
100 FIFTH AVENUE
NEW YORK, N.Y. 10011
WWW.ABRAMSBOOKS.COM

COLOPHON

Concept development,
design, typography,
photo research, and
production by
studio blue, Chicago.

Typeset in HTF Didot,
HTF Saracen, Monotype
Grotesque, ITC Bodoni,
ITC Franklin Gothic,
and Bell Gothic.

Color separations by
Professional Graphics Inc.,
Rockford, Illinois.

Printed by Cantz, Ostfildern,
Germany.